NATURAL CREATION &
THE FORMATIVE MIND

Born in 1944, John Davidson has had a lifelong interest in both the mystical and the scientific. Graduating from Cambridge University in 1966 with an honours degree in biological sciences, he took a post at the University's Department of Applied Mathematics and Theoretical Physics, where he worked for seventeen years.

Whilst studying for his degree, he had come into contact with the teachings of an Indian mystic and in October 1967 he made the first of many trips to India to meet him. He has been a follower of the mystical path since that time.

He left University in 1984 and presently runs the *Wholistic Research Company*, which supplies books and products related to living healthily in a modern world. Many of these products are of John Davidson's own design.

Since 1984, he has also written a total of ten books, including a series which attempts to demonstrate that *all* human experience lies within the framework of a greater mystic reality. His particular interest has been to show the true place (and value) of modern science in relationship to that reality.

NATURAL CREATION & THE FORMATIVE MIND

JOHN DAVIDSON
M.A. (Cantab)

ELEMENT

Shaftesbury Dorset ● Rockport Massachusetts

First published in Great Britain in 1991 by
Element Books Ltd
Longmead, Shaftesbury, Dorset, England.

First published in the USA in 1991 by
Element Inc
42 Broadway, Rockport, MA 01966

Designed by Roger Lightfoot
Front cover photograph: The Telegraph Colour
Library
Cover design by Max Fairbrother
Typeset by Selectmove Ltd, London
Printed and bound in Great Britain by
Dotesios Ltd, Trowbridge, Wiltshire

British Library Cataloguing in Publication Data
Davidson, John
   Natural creation and the formative mind.
   1. Cosmology (Metaphysics)
   I. Title
   113.8

ISBN 1–85230–197–X

CONTENTS

DEDICATION

To the Mystic Ocean of Being
In which we live

NATURAL CREATION
AN INTRODUCTION TO
THE SERIES

Natural Creation is a series of three books continuing the author's presentation of the world of science and humanity within a context of universal and natural mystic philosophy.

The three books in the present series are entitled:

Natural Creation and the Formative Mind
Natural Creation . . . Or Natural Selection?
Natural Creation: The Mystic Harmony

They may be read in any order, though they were written in the above sequence.

Natural Creation begins with the premise that this physical world is only a level of perception, a plane of consciousness, a dance of *continuous manifestation* or creation, a play of the Universal Life Force, a realm of being within the all-encompassing Supreme Being.

It points out how the Universal, Cosmic or *Formative* Mind is the architect of all patterns and rhythms, of space and time, demonstrating that physical science is presently only the study of superficial relationships upon the 'surface' of this Golden Womb of creative manifestation. It shows the way forward for man and his science to work in harmony with natural law. It suggests a radically different perspective on the true nature of science.

Natural Creation portrays the world of our fellow species as living and engaging beings, possessed of mind and consciousness. It shows how strictly Darwinian suggestions concerning evolution as an explanation of the fossil record, and of how we come to be here, cannot be the whole story since they deal only with bodies, ignoring the inner dimension of Life and the great creative power of the Formative Mind.

The author points out that present scientific and mechanistic descriptions of the universe are not so much incorrect as lacking in their perception of this inward formative dimension of being. They are therefore fundamentally incomplete.

He shows how everything is integrated and ordered into the most beautifully exact cycles and patterns, from atoms to galaxies; how the planetary surface constantly moves and recycles itself for the continuous support of living creatures. He presents a logical, yet mystical perspective of life on Earth throughout the millions of years of our planet's history. And he explains fully the tremendous formative impact of the most significant of all cycles upon planetary circumstances: the ebb and flow – with a periodicity of several million years – of Mind and Consciousness, of the Life Force itself. Evidence of this great cycle is clearly etched into the Earth itself, for all to read.

He directs attention to the evidence that, on a geological time-scale, we are nearing the 'end' of this cycle and may 'soon' experience a springtime of consciousness, the like of which we can hardly begin to conceive. He presents the mystical experience as the highest of all experience, showing how it is, and has been, common to all peoples and to all cultures. He ends the series with a full description of the mystic hierarchy of creation, according to the teachings of the highest mystics.

ACKNOWLEDGEMENTS

So many people have contributed to this book, in so many ways, that it would be impossible to mention all of them by name. But in particular, I have quoted from the writings of all those listed below. I am grateful to the authors, translators and publishers for the use of this material. Full details are given in the bibliography.

Ape Language, Sue Savage-Rumbaugh.
Autobiography of a Yogi, Paramhansa Yogananda.
Early Christian Mystics, edited and translated by A. Mingana.
   The quotation used is from Simon of Taibutheh.
The Imprisoned Splendour, Raynor C. Johnson.
Kinship with all Life, J. Allen Boone.
Man-Eaters of Kumaon, Jim Corbett.
The Master Answers, Maharaj Charan Singh Ji.
The Outermost House, Henry Beeston.
Paracelsus, Robert Browning.
The Philosophy of the Masters, Maharaj Sawan Singh Ji.
Some Problems of Echolocation in Cetaceans and
   The Echolocation of Marine Mammals, K.S. Norris.
Spiritual Gems, Maharaj Sawan Singh Ji.
The Story of My Heart, Richard Jeffries.
Supersense, BBC TV series.
Tuning in to Nature, Philip Callahan.
A Walk with a White Bushman, Sir Laurens van der Post.
Watcher on the Hills, Raynor C. Johnson.
The Way of Mysticism, Joseph James.
Whales, Nigel Bonner.

Also quoted are William Blake, Goethe, William James and Milton.

Thanks are also due to Christian O'Brien and Julia McCutchen who went through the manuscript making valuable suggestions, to Nancy Mangan who did much of the typing, and to Dennis Halls for the neatly executed drawings.

Awake!
Awake, O Sleeper in the Land of Shadows,
Wake! Expand!

William Blake

Truth is within ourselves; it takes no rise
From outward things, what'er you may believe.
There is an inmost centre in us all,
Where truth abides in fulness; and around,
Wall upon wall, the gross flesh hems it in,
This perfect, clear perception – which is truth.
A baffling and perverting carnal mesh
Binds it, and makes all error: and to know
Rather consists in opening out a way
Whence the imprisoned splendour may escape,
Than in effecting entry for a light
Supposed to be without.

Robert Browning (*Paracelsus*)

The hours when the mind is absorbed by beauty are the only hours when
we really live … These are the only hours that are not wasted – these
hours that absorb the soul and fill it with beauty. This is real life, and
all else is illusion.

Richard Jeffries (*The Story of My Heart*)

Sometimes I have concentrated myself, and driven away by continued will
all sense of outward appearances. Looking straight with the full power of
my mind inwards on myself, I find 'I' am there; an 'I' I do not wholly
understand, or know, something is there distinct from earth and timber,
from flesh and bones.

Recognising it, I feel on the margin of a life unknown, very near, almost
touching it; on the verge of powers which if I could grasp would give me
an immense breadth of existence, an ability to execute what I now only
conceive; most probably of far more than that.

Richard Jeffries (*The Story of my Heart*)

INTRODUCTION

One of my earliest loves was – and still is – natural history. The amazing intricacy of life holds as much interest for me now as it ever did. But the emphasis of my perceptions has changed. Whereas before I was enraptured by the beauty and detail of the physical form and the life stories of living creatures, now I am far more aware of the life and mind within them, of that which is responsible for the manifestation of the outer form. It is this expression of life and mind, the dynamic manifestation of the inward patternings that in their turn give rise to outward appearance, which is the major focus of the present book.

An understanding of ourselves and of nature is within the grasp of all of us. It is man's heritage to know himself, mystically, and to know thereby all else besides. For man is the microcosm, 'This Little Kingdom', within whom lie the keys to the greater empyrean. And as we travel on that mystic journey to the inward centre of the universe, we become aware of the deep harmony, balance and rhythm that lies at the heart of nature's processes, despite some of its more superficial appearances.

This world has never been an unalloyed paradise, though conditions here have been and will be better. Indeed, even now, the circumstances in which humans and other creatures find themselves vary widely. But at this point in our history, our complete embrace of science and technology has brought us to a place where we are capable of destroying and indeed have already destroyed vast areas of life upon our planet.

Such activity clearly reflects the condition of our own inward life. In no way can it be called advancement, for the only true potential for progress or evolution lies within the field of spiritual consciousness. Spiritually, we are presently in a degenerative phase, which will automatically be corrected by nature. A glimpse of this can be seen in the strong and rising undercurrent of universal spirituality.

But the process of correction, of learning the hard way, is always painful, though the result may be one of inward growth.

I hope that the understanding of nature expressed herein will contribute to the easing forward of this process of correction, making the path smoother. For if we could only learn to love all our fellow creatures as a reflection of our own understanding of life as love in manifestation, then we would treat our planet and all souls who live on it with the reverence due to the Creator, the Supreme Creative Life Force, who lives within us all.

In *Natural Creation and the Formative Mind*, I attempt to portray something of the inner life and mind of human beings and our fellow creatures. This is done from the experiential, rather than the intensely intellectual and analytical point of view.

Firstly, this entails an overview of nature and all living beings from a mystical standpoint. Mysticism being, in essence, the practical, experiential exploration of Being, under the driving and guiding power of Love.

This exploration leads us directly to the mystical understanding that this world is not a 'solid physical reality', but a Mind world. The panorama experienced through our physical senses, which we call the material world, as well as our individual minds and other aspects of our being, are all manifested patterns formed by a greater Mind the Universal or Formative Mind.

However, this breathtakingly integrated and beguiling pattern-maker is only an intermediary in the Divine Creative Process. The One Divine Source of Being lies at the inward centre of everything, while the dance of creation is spun out by this greater Mind. And the essence of our own being, our soul, is a drop of that Divine Ocean.

John Davidson
Cambridge, June 1990

1.   MYSTICISM AND THE NATURAL WORLD

NATURE'S WHOLENESS AND MAN'S DIVIDED MIND

We have an old and magnificent horse chestnut tree in our garden, towering high above the house. In a good year, the conkers come raining down in the autumn as the leaves turn from a rich green to yellows, browns and gold. It is a wonderful sight and the spiky chestnuts clatter through the foliage, falling with a satisfying thud upon the lawn, splitting open as they land to reveal the most beautiful, shining, pristine fruits within. It is one of my delights to collect them from time to time, carrying great handfuls to lay at the base of the tree for the squirrels or any other creatures who may care to take a meal from them.

Folklore always tells us that a heavy harvest of such fruits, hips, haws and seeds foretells of a hard winter, though it is not always so. It is nature's provision, say the wise old wives, for a hard-pressed community of creatures. We, with our modern, logical minds and our assumption that time progresses only linearly like a steam roller out of control, from past to future, find it difficult to perceive the whole of nature as one plan – a wholeness and an implicate order, where past, present and future are collapsed into one dance, into one mystic whole. It is not easy to realize that such conceptions of linear progression are more indicative of how our human, intellectual minds function than of how nature organizes herself.

The integration of relationships and rhythms within nature's magic matrix is beyond our limited, intellectual minds to fully grasp and to comprehend. The forces that shape our lives and our destinies, the coincidences and the patterns – are these all without rhyme or reason? Is there no greater consciousness than our own? There is. We are not left to drift so endlessly that only statistics and mathematics can tell us where we might be going.

This wholeness is everywhere and within everything. It starts within ourselves. We are the perceiver and experiencer of all of this. No academic institution or scientific dogma may call to account our inner experience of life and consciousness. Freedom is of the spirit, not even of our habit-ridden minds, and certainly not of the body, hemmed in on all sides by circumstances beyond our control.

We gain all by relinquishing all. By merging with the flow of life, by tuning into the patterns of our destiny and the rhythms of nature, we are able to rise above the constraining forces of our logical intellect. Then we see our life and all things with a more intuitive and inward eye. Hard times and good times are all balanced. Nature does provide and nature does deny, too. But there is far more to it than a simple linear cause and effect pathway, progressing blindly through time. There is a fabric of inter relationship, spun out, no doubt, under a law of causality whose ramifications we do not fully understand, but spread out over time, past and future. And this tapestry involves all the inward and outward aspects of our lives, as well as the patterns of the centuries and the millennia. The time-scale dates back millions upon millions of years.

We need to become aware of our innate faith. Faith that the same patterns will continue to revolve, shifting emphasis, but in essence the same. We assume or have faith in the constancy of the apparent laws of nature. We expect the sun to arise in the morning. We expect electricity, gravity and light to behave in the same way tomorrow and next year. We presume that our family, friends and associates will remain as humans and not in some grotesque Kafka-esque fashion, have become unrecognizable when we greet them in the morning. This much unconscious, innate faith we have. We only need to make it conscious, to live our lives as co-workers within the intrinsic, natural law or patterns of the Creator.

The American Indians and many others have called it living in the presence of the Great Spirit. Religious folk call it living in His Will. This we all do whether we know it or not, for our own apparent free-will lies within the boundaries of the greater Will. But by moving out from the confines of our egocentric limitations, we see, more and more, the Great Power at work and realize our own incapacity to alter the primal patterns of creation, even in the manifestation of one subatomic particle. We cannot actually create anything, we can only rearrange the energy patterns. And that, too, only according to the hidden law.

So autumn and summer bring a harvest of foods for all creatures to dine upon during the harder winter months. And yet that harvest for another's palate is the bodies of other living creatures. Can we say that a beneficent Creator has structured life in such a way that life can only be maintained by the destruction of other life? How can such a strange affair have come about?

The mystics have all said that this world is just a plane of consciousness, just a level of perception, that we do not see the whole story. In fact, they say that it is just an illusion. No life ever dies. The body dies, but the inward soul, and mind too, speed onwards, according to laws we do not, as humans, perceive in operation.

To gain this perception we must see nature as a whole, integrated network that includes the inward force of life and consciousness as its creator and maintainer. A linear, logical, tunnel vision cannot perceive this. For the strictly intellectual mind, cause comes after effect, like soldiers marching in a column. It does not consider itself to be integrated into a cosmic design and dance in which time itself is a part of creation and has no absolute reality.

Death is not the end of a soul. It is simply the moment for a change of garment. But responsibility for the death of other creatures brings us back here again and again. This is also a part of the intended pattern. This is why the highest yogic paths insist on a reverence and respect for life. Life must live on life in this physical domain and, therefore, to rise above it in mystic experience, the downward drag must be kept to a minimum. Hence the advice to take our food only from creatures of the vegetable kingdom, which suffer the least when the inward mind and soul are deprived of their temporary physical vehicle.

And this is a dictum based, not upon health considerations, but upon a deep, mystic perception of the way things are. But materialistic thought sees life as no more than a chance by-product of molecular processes. The subtle forces of mind function and the hidden power of the Life Force are thereby lost from view.

MAN AND NATURE

Modern, Western man has conventionally studied the life both of himself and of other species as if both he and they were comprised entirely of physical substance. He has only rarely asked himself the question, 'What does it feel like to be a plant? Or an insect? Or a fish, a bird, or a mammal?' This is because we are largely unaware of our own inner processes of consciousness, mind, emotion and body.

We do not ask ourselves the question, 'What am I?' and so we fail to ask the question, not only of the lower species, but even of our fellow human beings, of those with whom we live out our days.

In this way we live blindly and our expressions of understanding, in scientific and materialistic terms reflect that blindness. If you open a university course book on insects or any other creature, including man, it will describe to you what they look like – their nervous system, their nutritional, digestive and reproductive systems, and so on. It may describe their sense organs and maybe even some of their behavioural patterns, but rarely, if at all, will the author encourage you to try and understand what it might be like actually to be one of these creatures. Even behaviour is studied in an outward way: 'The creature does this and does that.' We are rarely invited to mentally put ourselves in their place, to try for a moment to experience life as they might experience it.

Medicine is taught in a similar fashion. How the patient feels and is experiencing his condition is often considered of far less importance than his symptomology. Even psychiatry, supposedly a study of the inward mind, is largely taught and practised by the outward behaviour pattern and the symptom.

Dating back largely to the time of Carl Jung (1875–1961), some of the more aware schools of modern psychology do attempt to get into the mind of the other person in an empathetic and caring fashion and this is a reflection of the expansion in consciousness and awareness which is evident at this time. But although there is an increasing impetus in this direction, we still have far to go.

In the natural world, our appreciation of the Life Force and being within every living creature is expanding. For this reason, many folk are instinctively seeking a vegetarian diet, though there is a curious facet of human nature that often makes a person deny that they have become so on any compassionate grounds. It is as if such an admission of noble fellow feeling and regard for life were something that might be laughed or even sneered at. A rising consciousness, however, automatically leads us to respect the life both within other creatures and in ourselves, and vegetarianism is just one of its natural outward expressions.

When I look through the window into my garden, I see a host of creatures – plants, birds, insects – even small mammals – squirrels, shrews, mice, voles and bats. By night, we have badgers and hedgehogs. My interest in such creatures has been lifelong and I studied zoology as one of my courses at university, as well as medicine. But I would account all of that knowledge, useful though some of it may be, as of greatly limited value, if I could know how a plant *feels* about being a plant – how it experiences itself. Or how an insect, a bird or a mammal feels within itself. Experience is always of greater value than intellectual knowledge. Experience breathes life into knowledge.

This summer, like the last, a family of squirrels have regularly come into the house for nuts. The mother squirrel, Knutkin, comes to the window ledge and bangs on the kitchen window when I come downstairs in the early morning, my mind far from squirrels and intent only upon the tea-pot.

She fixes me with her eyes and stands up on her hind legs, peering in. I understand her 'thought' and reach into one of the nut jars. She gets excited and jumps down, running to the outside kitchen door. When I open the door she peers in nervously and waits expectantly for her morning feed. I observe her as she feeds, her bright and liquid eye ever watchful. And I wonder at the nature of her mind or inward structure, for she has very clearly *learnt* all the details of her association with me. And though she trusts other people, too, it is myself with whom she has developed the deepest relationship. And I am considerate of her feelings and instinctive apprehensions and do not make loud noises or move suddenly. But she likes to hear me speak soothingly to her and will stay longer when friends come by if I speak to her, empathize with her, and let her know it is all okay.

But I do not know how she feels or experiences her life, what she feels when she comes on a wet day, her fur damp and bedraggled, or how she feels about the cold summer weather that is so often our lot in rainy England, even in these days of the Greenhouse Effect. In the spring, when she was pregnant and expecting her babies, she would bury some of the nuts I gave her. She had a definite sense of purpose as she dug little holes in my wild flower beds and carefully pushed a nut to the bottom, skilfully pawing and padding the earth back into place. Whether she ever found them again, I know not. And whenever I have been away for a weekend, or even a day, she is pleased to see me – no doubt for her easy meal, but I think, too, for more than that.

I am not so very sure whether I have trained the squirrel or the squirrel has trained me. For it is entirely her idea to jump on to the window ledge, standing up on her hind legs, forelegs and nose to the glass, fixing me with a look of intent until I get her a handful of peanuts. Let us say that the situation has evolved out of mutual self interest: she for food and perhaps companionship; I, too, for companionship, as well as interest in her inner state.

Her three children, whom she raised during the first year of our acquaintance, were delightful, though she soon lost the smaller one

– maybe to misadventure with a crow or cat. Everything was a source of delight for them as they played and learned about their new life. Every branch and twig was investigated, every plant was jumped at and on as they sniffed and whiskered their way along narrow overgrown pathways and through the wild flowers. But how did they feel? Their body language was expressive enough and told me much, but not what I wanted to know. What is it like to be a squirrel? Or a young squirrel?

Zoologists, and scientists generally, feel mostly that such questions are subjective, even anthropomorphic, and to be avoided. But to live in harmony with nature we must learn to tune in and to empathize. This is the least we can do. For without that, we are likely to take nature as the enemy and try to control, exploit, conquer and violate her. But we are a part of nature and without realizing it, we will automatically bring these very acts back upon our own heads. The balance of nature will always play the last card.

There is a blackbird, too, who comes for the currants he loves so well. During the summer, he must feed them to his mate sitting on her nest, and maybe to his offspring, too, for he stuffs them into his beak and carries them off, so until winter comes he is only given a few: there are better foods in spring and summer for him to find for himself! He sits in the tree when he sees me at my dining room table and fixes me with a look until I relent and go for the currant jar. Then he flies on to the flat roof just above the kitchen door and waits until I emerge, flying down to his feast as soon as I withdraw a few yards.

I called him Scruff because the first year of our acquaintance he seemed to have got himself into a near-lethal escapade losing most of his tail feathers and plenty from his wings. This, combined with the efforts of raising at least two families in the year, plus his summer moult, left him – come July and August – with the sure prize of being the worst-dressed bird in the garden. His spirit, however, remained undaunted, and he made up for lack of feathers by beating his wings twice as fast, looking for all the world like a miniature black helicopter as he whirred his way up the steep ascent from the lawn to our rooftop, from where he could command a good view of all the local proceedings.

We have a male chaffinch, too, who announces his presence with a loud, 'Chink, chink, chink', hopping about excitedly outside the glass kitchen door, darting pointed looks at me through the window until he gets his seeds or breadcrumbs. He, too, has a distinct personality and a way of being that is clearly different from Scruff or Knutkin. His name, of course, is Chink.

The robin, like the squirrel, comes into the house when I leave the door open, or he fixes me with beady eye from atop the clothes-line pole. Sometimes he (or is it a she?) flutters up and down the window to attract my attention and to convey his message. He knows I have some toast crumbs for him if only he can let me know he's there.

His perky bravery is sharply contrasted by a female blackbird who sits on the bird table in winter and tries to drive the other birds away, but yet is the first to flee, usually to skulk under a prostrate juniper nearby, when I open the kitchen door, though the sparrows and the tits care not one jot for my distant presence.

The plants, too, respond to care in more ways than the purely physical, as many folk know, and many an interesting seed has found its way into my garden and made its home there.

Harmony attracts harmony, life attracts life, and the subtle patterns and rhythms flow abundantly onwards. But how do they feel? Brother caterpillar and sister squirrel? What is their personal experience of being alive?

For the majority of human beings, the sum total of our experience is contained within our sensory perceptions, our outward activities and our mental and emotional cogitations thereon, both conscious and subconscious. To this, a mystic adds an expanded, inward experience of the nature, not only of his own mind, but also of more inward realms. It is fair to assume, therefore, that the same pertains in some degree to the lower species. Their awareness of their own life and existence is through their sensory input, their outward motor activity and their inward instinctive and learnt mental patterns, which we observe outwardly as their behaviour.

We cannot, with our physical senses, see the mind and inner life of our fellow humans. Yet we do not doubt that they are there. Can we doubt, therefore, that in their own way, other species also have a mind and an inner consciousness? But exactly how that experience is felt is difficult to say. We have a tendency, for example, to think that their vision might be similar to our vision, but this would seem to be unlikely. Take moths, for instance, and probably many other insects, too, who, in their sensory perceptions, are creatures of the electromagnetic energy fields.

The antennae of many moths are tuned, just like complex TV or microwave radar aerials, to pick up very specific infrared (and microwave) electromagnetic patterns through the structure of their tiny, but specifically designed fibrillae (see Figure 1.1). They can, thereby, 'see' specific aspects of heat radiations, the infrared emanations of all objects, emitted both during the day and at night, each with its own distinctive pattern. This includes the emissions of living plants and all other creatures. Even the air itself has a constant infrared glow, an 'air-glow'. By this means, they recognize fellow members of their own species, as well – it is thought – as their food plants and much else besides. All this is discussed later in greater detail.

Some birds, too, are sensitive to the Earth's magnetic grid and the light from the sun and stars. They use such awareness for navigation, for migration, and for finding their way to food or home. Some pigeons can see for thirty miles on a clear day, or hawks can see tiny prey – even beetles – from way up in the sky.

But what kind of a world do they perceive in this way? Moths' eyes see from the ultraviolet down into the area of the spectrum we humans perceive, while their antennae 'see' the lower ranges of infrared and microwave. Even their detection of sex pheromones, molecules emitted by many female insects, is electromagnetic, for their sensitive antennae are thought to pick up the modulated infrared, laser-like, coherent molecular emissions, rather than actually 'smelling' them in the manner familiar to ourselves. The males can thus see the females at a distance, because they can 'see' molecules, just as we would see a coloured gas, and this explains why moths will arrive from upwind of a female as well as downwind, and why pheromone concentrations can be so small for detection to take place.

In fact, many female moths remain in one spot, vibrating their wings, with the result that their thorax heats up and the consequent infrared emission, becoming visible above the background, is modulated by her wing beats – a personal, species-specific 'Welcome and at home' message to males passing by at a distance, in search of love.

But what does this actually mean in terms of their experience of life? For we perceive them with our five human senses, yet the way they perceive each other and themselves is clearly different to the way we do.

When you put out your hand to catch a butterfly, but it flies away just before you touch it, it is not seeing your hand in the way humans see it. Is it perhaps aware of the infrared emissions? Certainly, most butterflies only really fly when the sun is shining and their eyes can catch its ultraviolet light. So is it actually 'dark' for them when the sun goes in? Or is 'darkness' a purely subjective, human perception? And what does a bat perceive with its audio-sonar, or the whales and the dolphins? Or the electric eel that tunes in on the pulsed, electrical radar of the knife fish, or the shark which is sensitive to the electrical neurological activity of its prey? Or sea birds which can recognize the

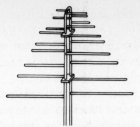

Array aerials, whose elements are on a mathematically defined scale of decreasing length, providing sensitivity across a wide band of frequencies

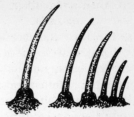

The fine hairs or sensilla on the antenna of the cabbage looper and corn earworm moths are arranged using the same mathematical scale and precision

Helical and equiangular spiral aerials, also built to a mathematical precision

Conical antennae are found on the Florida scarab mite. Many species of wasp possess thin, tapered, helical sensilla

Horn configuration, microwave aerials

'Shoehorn' and 'eared' furrowed sensilla are found on the antennae of many moths

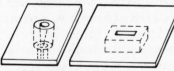

Cavity and slot aerial wave guides

Various kinds of sensory pits and cavities are found on the antennae of bees, wasps and ants

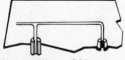

Loop aerials, useful in establishing direction

Loops are found on the antennae of many species in the Cecidomyiidae family of flies

Figure 1.1   Man-made electromagnetic aerials and their insect counterparts. Adapted from Tuning in to Nature *by Philip Callahan.*

voice of their own offspring even amidst the clamour of a colony numbering thousands?

Every creature has its own tale to tell. Each one lives in its own world. Clearly a world of perception and instinct which is vastly different to ours, and is yet linked to it, almost in the nature of overlapping dimensions. For when we perceived any creature, we perceive it through our five senses and react to it accordingly, according to our way of being and behaviour. But how they live and feel is quite different.

In this book, I would like to indicate a way in which we can begin to glimpse this inner world, for I would suggest very strongly that it is the life within any creature that is vastly more important than its anatomy and physiology. And it is this life which needs to be understood if we are to live in harmony with nature.

Modern man has fought with nature. He has never lived in harmony with her. But from this relationship there is no divorce, no escape. We are bound inextricably to the circumstances of our own creation.

Our culture in its present condition is not suitable for continued, long-term, survival. It is too destructive, and is already destroying itself and its habitat. So we have a great task ahead of us to bring our technology and social structure into harmony with our planet and with the life within both ourselves and within every creature. Many folk realize this, but many do not, and it is essential that this greater understanding is given practical shape at governmental and industrial levels, before the inevitable backlash of nature can be sublimated, redirected, dissipitated or dissolved from within itself.

So we need to rethink our activities upon this planet and redirect our energies towards more worthy goals. And although much can be done in a purely pragmatic and practical way, for those who are more spiritually inclined, it is of value to approach life and its problems with a comprehension of the inward mystic structure of the cosmos and thence of ourselves and the lower creatures.

MIND: THE SOUL'S SPACE SUIT

I have, in my previous books, *Subtle Energy* and *The Web of Life*, attempted to give a more complete and detailed picture of the mystic reality, inasmuch as words may describe what is really to be experienced within. However, for the purposes of this book, I will here give a brief resumé.

All true mystics have agreed that the essence of life is soul, or consciousness. That all living creatures, whether plants, bacteria, dinosaurs, shellfish, animals, birds or men are, in essence, drops in the Divine Ocean of Being, Consciousness or Life. The material universe, too, is His creation, His play or projection, and there is no part in which He is not present. He is within everything, there is nothing in which He is not the inward and continuously active Creator. He is within everything, and yet he is also aloof.

This creation, however, consists of far more than just our physically manifested universe. There is, say the mystics, a hierarchy of inner worlds of increasingly fine or subtle vibration, of which the physical universe is only a reflection. Thus, what pertains here has analogues in the higher or more inward realms. 'As above, so below'. This ancient Hermetic axiom is intrinsic , for the outer is a creation from the inner and bears the same characteristics.

Around the region of the Source, the One, the Universal Consciousness, in which there is no duality or differentiation, there are the purely spiritual realms, where souls are separated only by the Will of the Supreme. Descending further, however, the soul or drop of the Ocean of Life or Consciousness, meets the region of the Universal Mind. Here, duality first becomes explicitly organized, precipitated or polarized, and here matter and mind, as we know them in the gross physical world, first manifest in forms which we may recognize, though of an infinitely more subtle nature.

Mind, being the originator of explicit division and differentiation, is also the origin of time and space, the dimensions within which change and difference are expressed and experienced. For the soul, therefore, the Mind is a real *space suit*, for Mind is the creator of space.

Above the realm of the Universal Mind, the supreme law is that of Love. Below that, the law is of causality, of justice, of Oneness perceived through the veil of manyness. All the natural laws that man feels he has discovered are intuitive or intellectual reflections of our physical mind and thought, discerning limited aspects of this great and simple law of causality, polarity and differentiation. The Oneness is perceived through the web of the Mind, of the many, and is seen only as relationships within the web. The relationships we call causality, yet we do not fully understand its origins. But it is only the One Light, split into a myriad fragments. When we lose sight of that One, then we get lost in our attempt to understand the nature of the web, and how it has arisen.

This law of the Mind, of relationship, of causality, of *karma*, governs the energetic movements not only within the realm of physical substance, but also within the finer vibrations of the matter, energy or substance constituting our own individual minds. What we perceive as the energies of this world – physical or mental – are thus no more than the power of the One, split or divided by the Mind. Ultimately, the One is both the Primal Energy and the Primal Consciousness or Being. Everything arises within that wholeness. So energy as we know it, is only the patterns spun across the face of the Universal Consciousness by the greater Mind, in all its aspects. And our human mind centre is just one small part of that greater or Universal Mind.

Within this realm of the Universal Mind, then, first originate the ultra-fine subtle mental essences, energies or blueprints which – after further outward projection or emanation – we perceive on the physical plane. Just as we can appreciate space in our *mind*, as well as perceive it with our *senses*, so too do these finer essences originate in the higher realms of the Mind. They are the mental blueprint which is reflected down through the layers of being until we perceive them at what we call the physical level, but which is, in reality, only a level of the Mind.

These essences or modes of vibration, are known to Indian mystics as the *tattwas*. They are – in ascending order – earth or the solid state, water or the liquid state, fire – which drives the transmutation between these states, taking matter into the plasmic condition and includes electromagnetic radiation, air or the gaseous state, and *akash* – the locked-in, energetic potential of the vacuum state. Akash or vacuum is the more subtle creative matrix or space from within which the other tattwas or elements dance and spin into existence, under the law of cause and effect operating both vertically or from within-out, as well as in the horizontal manner with which we are accustomed.

These elements or tattwas, the five basic vibrational octaves of material or energetic substance, are discussed more fully in *The Web of Life* and the akashic or vacuum state is discussed in some detail in *The Secret of the Creative Vacuum*. However, it must be emphasized that the tattwas as we experience them physically are only reflections of reflections, of what lies within. First originating in the exquisitely subtle realms of the Universal Mind, they are reflected downward, through the astral realm, until we finally perceive them as the substance of our physical universe. Like the sun which is reflected in a bucket of water and thence upon a wall, the original clarity or truer reality is increasingly obscured with each reflection. And at the human level we are lost in ignorance, taking the diffused reflection upon the wall to be the only reality.

Now it must be understood that these five octaves or tattwas comprise *all* the 'substance' that there is, both in the physical, as well as in the more subtle and inward realms. This means that not only does our physical universe manifestly consist of these five tattwas, but our inward mental, sensory, behavioural and psychological processes also consist of them in their various subtle forms.

And it is variations within this subtle tattvic tapestry which give rise, not only to the differences of personality, talent, skill and psychological characteristic with which we are familiar both within ourselves and our fellow humans, but also to the differences in our outward physical characteristics, too. For the inward and the outward are both part of the same integrated energy complex which we observe as a human being. The outward is actually derived from the inner.

We can get some glimmering of how this physical world is actually an aspect of the greater Mind when we realize that what we call the physical universe consists entirely of subjective sensory experiences within our own apparently individual minds. Even our sensory experiences are within ourselves – subjective. For example, we cannot even convey to others our *experience* of colour. Our physical senses, then, are really aspects of our minds, as is the physical universe. And the very fact that we all perceive the physical universe – in our minds – demonstrates that there is a great unifying power within which we all 'exist and have our being'.

So all apparently objective outer experience (i.e. sensory perception and motor action) is actually entirely subjective and experienced within our own mind. And the same is true for our thoughts and emotional reactions, conscious and subconscious, concerning these sensory and motor experiences. It is all, indeed, an illusion, played out under the forces and our own *personal karma*, the impressions and residue of actions, thoughts and desires of previous lives etched into the fabric of our own mind structure and now played back to us as the destiny of our present life. The present is due entirely to the past, and in the process of living out the reaction, we create the seeds for the future.

And not only that, for the nature of all the other species both in their inward being, their mental-instinctive structuring, their resulting behaviour and their outward form, is seen to be only variations in the constitution of this inward subtle, tattvic tapestry. This also arises as a result of their own karma.

The elucidation of this subtle tapestry in the lower creatures is another of the focuses in this book. For it can be quite readily comprehended intellectually even without the mystic or inward vision that would permit us to observe what it is really like to be, not only another species, but also one of our fellow human beings – those whom we have a tendency to exploit, manipulate and generally treat so badly.

THE EGG OF BRAHM

It is difficult to imagine or convey the extent of integration within the processes of the Universal Mind, the great weaver of pattern and form. It is more holistic, holographic and integrated than we could ever imagine. In yogic writings, this Universal Mind is spoken of as *Brahmanda*, whose power or lord is known as *Brahm* or *Kal*.

Within the totality of Mind lie the *Three Worlds of the Mind* – the causal (*Brahmanda*), the astral (*Anda*) and the physical (*Pinda*). Everything within them, all forms and patterns, are thus of the Mind. And all forms are integrated and related to each other in one whole, powered by the inner Life Force, the Universal Consciousness, the Divine Ocean or God.

The word Anda, referring to the astral region, actually means *Egg*. It is for this reason that the Universal Mind, Brahmanda is often referred to as the *Egg of Brahm*, because an egg signifies a complete whole in which all its parts are integral in forming the nature and functioning of that whole. Furthermore, if you are a part of such a whole, it is incredibly difficult to break free, to rise above the completeness of a web which is so all-embracing that every aspect of one's being and experience is caught up in its operations. Even if we reach the 'top' and sit on the 'outside', we are likely to glissade down to the bottom once again – the contours of an egg make it an impossible shape to climb. So the metaphor is strikingly apt, as all those know who have struggled with any form of personal mind control or meditation.

But ultimately, there is only one Creative Power, only one Source. All *energy*, therefore, is a derivative of this central, uncreated Power. We perceive this energy as related patterns, spun across the face of the One by the machinations of the Mind. The Universal Mind, in any of its myriad aspects is the great weaver of form, pattern and relationship. And it is this creative dance of the One divided into the many by the operations of Mind, which we call energy.

It is true, therefore, that matter and energy are equatable, for matter is just Mind patterns, but our understanding of energy needs to be pushed a little further than present conventional scientific paradigms permit! For everything in creation is energy, *per se.* It is nothing but the One in patterned form – the Lord in a see-through, multi-patterned T-shirt! The T-shirt is the Universal Mind and, in a most mystical way, we are both a part of the pattern as well as the whole caboodle itself.

Therefore, the implicate wholeness of physical nature which we intuitively perceive to be in operation, is no more than a reflection or a part of the greater wholeness of the Universal Mind. For the physical universe is only one part, one layer or one dimension of the greater whole. This may be difficult to understand or believe. But bearing it in mind, at least as a possible working hypothesis, let us proceed!

THE GOLDEN WOMB

It is in the Vedas and other yogic writings, that the Universal Mind is called *Brahm*. It is also described as *Hiranya garbha*. In literal translation this means, 'One who has gold inside the womb'. The One is God, and the descriptive reference is to the transcendent light of this inner stage which shines with a golden-reddy colour.

It is called the womb because the Universal Mind is the womb of creation for the three worlds of the Mind. It could also be translated as the *Golden Womb*. Just as an embryo is formed and patterned in the mother's womb and out of the mother's substance, so are the three worlds formed out of Mind substance and patterned by the power of the Universal Mind, a power derived from the Supreme Being Himself. In the Upanishads, this formative region is also described as the *Golden Sun*, our earthly sun being the source of physical energy for life on Earth.

Is is these descriptions which are therefore the origin of the expression, the *Golden Egg*, Brahmanda or the Egg of Brahm, for an egg is an embryo within which, from within the heart of the golden yolk, an amazing transformation takes place. The material is repatterned, until a tiny chicken emerges. It is a totally internal process – nothing is drawn in from the outside. Surely there is a mystery here beyond the realm of self-organizing molecules? There is a Holy Ghost in the machinery!

So the Universal Mind is the Golden Egg, the Golden Womb – the Womb of all lower wombs, the primal pattern-maker whence all other patterns are derived. Similarly, the Universal Mind is also described as the Divine Mother, though this epithet is also given to the Supreme Being. And we also talk of Mother Earth, the planetary womb from which all physical bodies derive their substance. But the pattern-maker lies deeper than our outward human perception of substance. It lies in the formative character of the Mind itself.

The appreciation of how patterns come into existence in the creation gives us an understanding of how we come to intellectually appreciate logic, form, pattern and rhythm. Our personal mind, being a part of the greater Mind, unconsciously reads the patterns of its own inward nature and structure – both within itself as well as in the physical universe it considers to be without. Thus do we derive what we call the 'laws of nature'. But it is only mind reading Mind! For nature itself is only another word for Mind.

In science, geometry and mathematics are thus, in essence, only a language or a terminology for the expression of relationships in space and time. Space and time being fundamental attributes of the Universal or Formative Mind, it is easy to see how some mathematicians feel that mathematics is a fundamental attribute of nature. Actually, the relationships within the many do indeed make up that of which nature consists, but the *concepts* of mathematics lie within man's intellectual mind, as does all human language. So one can say that mind is thus perceiving Mind, the small part is looking at the whole and automatically finds an affinity between itself, its own manner of functioning, and that of the universe, 'outside'.

The structure of this whole Egg of Manyness, this great womb of pattern forming, is thus partially expressible by geometry and mathematics, as indeed it is by any terminology or language which describes relationships. But to reach its most powerful and truthful expression, the geometry and mathematics really need to describe the way in which the One becomes the many in the multireflective, multidimensional, multilevel, multifaceted, integrated worlds of the Mind.

Statistics, too, and 'laws' of chance are also only an appreciation of these relationships within the one integrated web of what we call natural processes. Our mathematically expressed 'laws of nature' appreciate some of the associations without comprehending their intrinsic reality, while our use of statistics indicates that we can see a pattern, but have no real idea of why the pattern should be there or how it has arisen. Statistical mathematics and 'deterministic' mathematics are thus both very similar. They are both appreciations of rhythm and pattern but without a full comprehension of how such order arises.

But the subject of form, pattern and rhythm is a wide one, discussed more fully in *Natural Creation: The Mystic Harmony*.

TERMINOLOGY OF THE MIND

It might be helpful at this point to summarize and clarify the meaning behind some of the terms used herein to describe Mind function.

1. *Universal Mind* refers to the first appearance of Mind as the soul 'descends' or moves 'outward' from its source. The region here created is the causal region. It is the primary pattern for all lower Mind activity. Space, time, polarity and 'substance' first arise here.

2. The *individual mind* of a creature, human or otherwise is written as mind with a small 'm'. This is sometimes called the individual *mind structure*, while the term *mind set* refers to the personal aspects and differences to be found even within the same basic mind structure which makes a particular species what it is. In humans, this is something akin to *personality*. Or rather, personality is one expression of our personal mind set.

    Emotions, thoughts, feelings, intuitions, personality, intellect, memory – all these and everything else which goes into making us feel like and behave as individuals – are all a part of our individual human mind.

3. When referring to the totality of Mind function, of which individual minds are just a part, then Mind is spelt with a capital 'M'. To be consistent, 'life' and 'consciousness' should also be spelt with small and capital L's and C's under similar circumstances, but the use of too many proper names would only create mild confusion. Mind, with a capital 'M', is sometimes referred to as the *greater Mind*, or the *Cosmic Mind*.

    Carl Jung's concept of the *Collective Unconscious* is an aspect of this greater Mind, though his thinking was more abstract and conceptual. We are dealing with Mind as an 'entity' that has an even greater reality than the 'solid' ground beneath us. It is something which one can, (figuratively speaking), stick a fork into or put in a jam jar.

4. The term *Formative Mind* is an expression introduced and discussed at length in Chapters 8 and 9. It refers to the fact that Mind is the former of all patterns, of all diversity, of all

actions and activity, of all multiplicity in our sensory world, as well as in personal experience of our individual mind. The Formative Mind really means the same as Mind, or greater Mind, but the term emphasizes the formative role of Mind in the creation. It therefore includes the astral and causal realms as well.

INTERLUDE

THE GARMENT OF GOD

Nature is but the living, visible, garment of God.
(Goethe)

What if earth be but the shadow of heaven?
(Milton)

In this world,
Anyone who does not believe in miracles –
Is not a realist.
(Anon)

2. LIFE FORMS AND THE FIVE ELEMENTS

THE ESSENTIAL ELEMENTS

In my book, *The Web of Life*, I expanded at some length on the subject of the five tattwas and ten *indriyas* and this topic was once again introduced in the previous chapter. As described, the tattwas or 'elements' are the essence of all material substance. Material substance, however, cannot be separated from the mind experience and senses of the living creatures perceiving it. Physical things are to us just what we perceive or experience of them. They have no other reality. The physical world is a subjective sensory experience in our minds. And all creatures perceive things differently, too.

By some inward mystic process, the souls are brought into the realm of the five physical tattwas, due to the tendencies, attachments and desires of their individual minds. Furthermore, the processes of creation are deeply integrated, for the very existence of the energy patterns which comprise the physical realm and which we experience through our mind and senses is dependent upon the hidden, inner linkage of the individual minds and senses of all the creatures involved. It is a part of the implicate integration of Universal Mind function. We are thus all co-creators or shareholders in the manifestation of our apparently outward sensory experiences, of what we think of as our physical universe.

One can understand this simply by means of an analogy. In this world, a business exists because of its shareholders. To a superficial eye, it appears that the business exists independently of its shareholders. But remove the shareholders and the business evaporates. It never really existed separately from the people involved. It was the people, particularly the *minds* of the people, which gave it the illusion of existence (for the sake of the example, one can consider all those involved with the business as 'shareholders', of one kind or another).

Similarly, all physically incarnate souls are shareholders in this world. The shareholding is the attachment, entanglement and activities of all their minds with the 'world'. So the 'world' only exists because of the souls and their minds.

When we die, we are forced to relinquish our shareholding in the world – but only temporarily. For our involvement is such that we are almost immediately attracted, like a magnet, to take up another shareholding via the process we call birth. The illusion of the world as a continuous and 'real' business is thus maintained since all the souls are continuously involved with each other, through their minds and through the medium of physical substance.

Now the five tattwas are vibrating essences of mind energy around the soul at whatever level, within the Mind regions, it finds itself. As human beings, we experience these tattwas in their subtle aspects as the five sensory (yin) and the five motor (yang) indriyas. It is also experience, at a subtle level, of these five tattwas which underlies our natural human faculties, as well as our emotional and psychological characteristics. Being human is actually a subtle as well as gross experience of perception and action through these five inter-related vibrational conditions of substance.

(Perhaps it should be noted that the Hindi word *indriya* literally refers to the five senses and the five primary motor functions. Each sensory or motor function is comprised of the physical organ and its mental counterpart. The term *indriya* is actually used to refer to both of these, but in my writing I have used it to refer only to the mental counterpart, in order to underline the differentiation between the two.)

As previously described, when we use our senses to perceive something, the *experience* of that perception is actually felt in our *mind*. In fact, when we close our eyes we can still see things *in our mind's eye*. Similarly, we can hear sounds or sing songs inside our own heads. Or we can recall smells, tastes and sensations without the physical sensory apparatus being active. When we do perceive through our sense organs, we use this same inward mental faculty as a part of the inward process of experience. This mental faculty is known as a *sensory indriya* and there are five of them, relating to our five outward senses.

And these sensory indriyas arise because the mind is moving outward and playing against or interacting with the subtle physical essence of the five tattwas. Five modes of sensory perception are thus formed in our human, subtle energy blueprint, which then manifest outwardly as our five senses. Each sense is therefore an expression of one particular

tattwa or inner subtle vibration, or condition of matter. In man, they
are:

Tattwa	Sense Indriya (Mental)	Sense Organ (Physical)
Earth	Smell	Nose
Water	Taste	Tongue
Fire	Sight	Eyes
Air	Touch	Skin, especially the hands and fingers
Akash	Hearing	Ears

To a Western mind, unfamiliar with the more inward Eastern idiom,
this may take some getting used to, in order that the meaning may be
grasped. But it is deep and fundamental. The subject is discussed at
length in *The Web of Life* and there is no need to go once again into
such detail. Some information, however, as it relates to the present
theme, does need a little elaboration here.

Actually, 'Earth', 'Water', 'Fire' and so on, are poor translations
because of their association, in our thinking, with only the physical
conditions. But, as we have seen, even these apparently objective states
are really subjective mental experiences, gained through our physical
senses, but experienced within ourselves. One cannot, for example,
convey one's actual *experience* of the colour red or the wetness of
water or the solidity of earth to anyone else. And since all sensory
experience is of this nature, the entire physical universe turns out to be a
quite incommunicable, subjective experience. Language only works as
a means of communication because we have all had similar experiences
and have agreed a certain noise (the word 'red', for example) to direct
another person's attention to their memory of a similar experience. But
if you do not speak my language then no amount of my shouting the
word 'red' will give you any idea of my inward meaning.

Similarly, one cannot dissect a human brain *and find any experience
at all* – thought, emotion or sensory perception. Neither will one find
the appreciation of beauty, or discover the nature of the mind or soul.
These are all at more subtle levels of energy and being.

So bearing this in mind, one can begin to understand how the eyes
are an expression of the subtle fiery tattwa. Light is electromagnetic in
nature. It is the fastest moving, expanding energy known to physical
science. This is a quality or aspect of the fiery tattwa and we experience
it as the mental indriya of sight which underlies the sensory perceptive
processes of our physical eyes.

It is very important to grasp this differentiation between the mental experience of the sensible world, and the physical senses themselves. When we use our eyes, we know nothing of our cornea, our variable focal length lens or the musculature of our iris which surrounds the pupil. We are quite unaware of the retina with its array of rods and cones, its electrobiological activity, the optic nerve impulses and our brain's activity. Analysis of these things leaves us with no comprehension at all of our actual *experience* of sight. An experience that itself is so inward that we can never share it with others, other than by the indirect and much imperfect means of language. We cannot convey to a man blind from birth the beauty of a sunset or even what we mean by the word 'red'.

But actually, on a day-by-day, second-by-second basis, it is this subjective *experience* through our senses and in our mind and emotions that is of the greatest relevance to us. This is how we *experience* our life, our consciousness, our awareness. Words and descriptions, however detailed and scientific, tell us nothing of our *experience*. They may help us to sort it out in our own minds, but description is no experience. You may take away all the books, all intellectual knowledge, but we do not wish to be deprived of our life, our existence, our experience of being.

And this fact, that our *life* – our being – is something deeper than our intellectual or philosophical analysis of its outward manifestations, is actually the factor which has been omitted from almost all of conventional science. For this reason, modern science is largely upon the wrong track, for it totally ignores the very life and experience that gives it existence! The Life Force within, the being within – of man and of all other creatures – is ignored. Hence the dreadful environmental mess created by man's technology upon this planet. Bio-logy – the study of life – does not, as it is presently practised, study life so much as it studies form.

Returning then to our discussion of the indriyas, the mental sense indriyas and the physical sense organs, together with the brain, handle the receptive 'input' from the physical world. Completing the circuit, our response, again through the tattvic energy fields (there is no other material substance, subtle or gross) is one of motor action. These *motor indriyas* are again mental faculties or intentions that manifest outwardly as:

Tattwa	Motor Indriya (Mental)	Motor Organ (Physical)
Earth	Elimination	Rectum
Water	Procreation	Sex organs
Fire	Going places	Legs, in particular
Air	Manipulation	Hands and fingers in particular
Akash	Speech	Throat and mouth

Again taking 'fire' as our example, note how this moving and expanding principle becomes expressed as walking, moving about or 'going places'. Normally this is accomplished with the legs, but even in their absence, the mental idea or ability to conceive the possibility of going somewhere remains, and may be expressed through the use of the arms or even the torso alone, if the arms and legs are tied or are otherwise unable to perform their normal functions.

All creatures except plants possess this inward 'mental' pattern which leads them to move about. Some may not move about very much, but then some humans are pretty sedentary, too! It is inward integration of the fire tattwa into the subtle blueprint of a creature which forms the corresponding mental, action indriya of fire resulting in all creatures other than plants being *able* to move about and *wanting* to go places. It gives them the *idea* that they can. Then the action, and the necessary anatomy and physiology, follow in consequence, as an outworking of the inner blueprint. This begins to tell us how a mind and body are put together.

Plants do not possess this faculty or inward integration of the fiery tattwa and hence they do not get up and walk. Interestingly, moving about requires a higher intelligence than just staying put, because the creature has to decide *where* to go, *what* to do, and it has to see *where* it is going and *how* to achieve what it has in mind. It also needs the capacity to learn about its local environment. Even the science-fiction author, John Wyndham, found it necessary to endow his famous walking plants, the Triffids, with both a limited intelligence as well as some degree of communicative ability! To move about requires the mental ability, however spontaneous and instinctive, to decide *where* to go and *what* to do. For this, a higher level of mind function is required than is provided purely by the one tattwa of water. This is the essential difference between plants and other creatures. Their inner, subtle structuring is different, and the outer form automatically follows suit.

Similarly, in man, the motor indriyas of air is expressed physically through the intricate manipulative abilities of the hands and fingers. We can physically arrange and rearrange things with our feet and legs or even with other parts of our body if need be, but it is the extension levers of the arms and the wonderfully intricate manipulative capabilities of the fingers which are our primary manipulative organs. The Latin word *manus*, of course, means hand, as in the word 'manual'.

Among birds, the airy tattwa is again expressed through the forelimbs, this time modified as one of nature's most intricate of mechanisms – wings to fly.

So we have five sensory and five motor modalities for experiencing the world of the five physical tattwas. And as we have seen, these possess both subtle or mental as well as gross physical aspects, manifesting in one whole integrated system of mind and body. The physical body is thus derived automatically as a part of the outworking of the inner mind energies. This is how there is such integration between mind and body. The one is actually derived from the other.

Similarly, our human psychology and behaviour is influenced by the characteristics of the subtle tattwas comprising the subtle energy blueprint. Man has physical and mental modes of expression which arise directly from the characteristics of these subtle tattvic energy fields, though it is difficult to find concise words to summarize it.

This difficulty is partly because man's understanding of his psychology has become so tortuous that we have largely lost sight of its more simple energetic basis. Carl Jung, for example, wrote a multitude of books attempting to express the complexities of both the balanced and the disordered human mind with which we all have to contend. Yet he never reached the end of his thoughts. Table 2.1, however, is a simple table attempting to express this inner constitution.

See how the expanding fiery tattwa provides the mental-emotional *drive* to accomplish things in the world. Or how the mental intricacy of air gives us the detailed imagination to arrange and rearrange the energy patterns of our physical existence. This is the quality which manifests as a bird's instinct to build such intricate nests or a spider its web or a bee its honeycomb or, when combined with fire, gives a bird both the drive and the ability to navigate over thousands of miles of previously unflown territory.

When we have understood something, intellectually, we even say that we have 'grasped' it. The airy tattwa is expressed in the same manner both in our hands as well as in our minds. And when out of balance, grasping, of course, becomes greed, both for physical and

Table 2.1 The tattwas expressed as human faculties and weaknesses

Tattwa	Mental Characteristic	Physical Characteristic	Imbalanced Emotion
Earth	Practicality, groundedness	Possession	Attachment
Water	Sensitivity, feeling	Procreation	Lust
Fire	Drive	Going places	Anger
Air	Planning	Procurement	Greed and grasping
Akash	Discrimination; perception of one's place in the scheme of things; identity	Executive capability	Egotism and pride

Note how the faculties, like the sense organs and their mental indriyas, have both mental and physical aspects. Similarly, their imbalanced condition, which constitutes our human weaknesses, have both physical and mental aspects. There can never be an outward action without a corresponding inner motive. Mental desire for possessions leads to their physical accumulation. Mental desire for sex leads to its outward expression. Anger is never expressed through word and deed without it first being felt within the mind. Similarly with greed and egotism. Weakness can be entirely mental. It may be suppressed – never expressed in action in any explicitly obvious way. Yet it subconsciously controls the entire being and activity of the individual. But it can never be present in action without first being present within the mind.

mental things. The insatiable desire to acquire material possessions over and above the needs indicated by the natural flow of our life is greed for physical objects. While the use of the intellect to ravenously acquire 'points' from here and there in order, later, to display our cleverness and knowledge to others is really no more than greed for intellectual or mental ideas.

Even when in balance and in the full consciousness possessed only by a perfect mystic adept, man still uses all these faculties for his expression upon the physical plane as a perfect human being. It is thus an instinctive and necessary part of being human to appreciate the need for *owning* various basic needs (possession/earth). We do not want to have to find new clothes and a fresh dwelling, every day and night. We also need the means for and to know how to propagate new bodies (procreation/water); how to achieve our ends (drive/going/fire); to get hold of and manipulate the things we need (procurement/air), and to understand our place in the panorama of existence (human identity/discrimination/akash). These things are all a part of being in a physical body.

But in most of us, our soul is under the sway of the mind whose attention runs outward to the senses and we thus develop a vast inward area of unconsciousness, our subconscious mind, in which – largely unknown to us – the imbalanced aspects of the five tattwas hold sway. Often we are so confused and unaware of what is going on within us that we take egocentric credit for what are really imbalances or weaknesses in our expression as humans.

In this way, our groundedness and ability to deal with and comprehend the grossly physical aspects of life become perverted into undue *attachment* to our worldly possessions, or even to our mental ideas, habits, social customs and religious rituals and beliefs. Similarly, our sexual proclivities, required for the propagation of the species and to give us a feeling of human companionship, become a personal self-indulgence and sensory delight or *lust*. Frustrated drive leads us to *anger*. The mental ability to organize, arrange and acquire things overbalances, leading us into *greed*, miserliness, and manipulation of other people in order to acquire more than we need and to achieve our own selfish ends. Akash, also reflecting as the vacuum state out of which all physical substance arises, become *ego* or self-identification, the point in our psychology from which all the other weaknesses are derived. Then we lose our sense of discrimination, of the ability to understand our place within the karmic skein of life.

But all these faculties and characteristics are woven together into one fabric. Just as all aspects of the three worlds are woven into one whole as the Egg of Brahm, the total structure of Universal Mind, so is man a whole – a reflection of a higher and more inward and all-encompassing totality. So all our faculties are interlinked. Underlying possession, procreation, drive and procurement is our sense of human individuality or ego. Possession, procreation and procurement presuppose the existence of the drive to act and to move. To *possess* things, we must first *procure* them, and so on.

When, as in most of us, the faculties essential for the continuation of physical life become unbalanced and we lose a clear vision of what we are really about, then the weaknesses are also merged. Sexual desire and the need for procreation lead to intense attachment to our partner, to anger or hatred if that drive is frustrated, to envy – a mixture of greed, attachment and egotism, perhaps – if someone else gets (procures) what we had in mind – and so on.

But expressions of these five basic energy fields cover all aspects of our psychological experience and behaviour. The simplicity of it is beautiful, and it is also true, because it is based upon the actual, inward energetic constitution of man.

Why are there five – not three or seven or some other number? Well, why do we have five fingers on each hand, two legs and one heart? The real answer to just one of these must ultimately be the answer to the others. But for that answer we need mystic experience. To see from within the manner by which the One becomes the many.

This simple energetic and experiential basis to psychology is also the reality underlying the science of astrology. Astrology – when it is understood and practised correctly – is a true science by which patterns can be read, concerning the inward nature and karmic expression (i.e. destiny, personality and so on) of individuals. The universe is like a hologram in which every part, or every pattern, can be used to project a pattern of the whole. Similarly, palmistry and numerology are sciences based upon the same pattern-finding principle. These days, of course, there are thousands of party astrologers and palmists, and very few people who really understand the science deeply. But that does not mean that the true science of astrology does not exist.

From the foregoing, it should have become clear that the solid aspect of earth (for example) is only one aspect of the manner in which we actually *experience* the tattwa of earth. Similarly for the other tattwas – the liquid, fiery, gaseous and vacuum states are only single aspects of the ways in which we experience the tattwas. But because our attention is outward, and our mind and senses are completely gripped by what we think of as gross physical matter, we tend to be aware only of this one outward aspect. Ignorance of a process, however, does not preclude its existence. Note, too, that the word ignorance means just that – to ignore, to direct our attention away from something, to overlook. But the understanding is there within all of us, if we care to look.

We are, then, in a position to draw up a table of these characteristics of the tattwas as we experience them in gross and subtle ways, (see Table 2.2) But remember that although this table is laid out in linear fashion, the human *experience* is of one integrated system of energy and being. And that words are only a pointer to reality.

Table 2.2 Aspects of the five tattwas in human experience

Tattwa	Outward form	Sense indriya (mental)	Sense organ (physical)	Motor indriya (mental)	Motor organ (physical)	Mental charac-teristic	Physical emotion	Imbalanced
Earth	Solids	Smell	Nose	Elimination	Rectum	Practicality grounded-ness	Posses-sion	Attachment
Water	Liquid	Taste	Tongue	Procreation	Sex organs	Sensitivity feeling	Procrea-tion	Lust
Fire	Fire & light	Sight	Eyes	Going places	Legs	Drive	Going places	Anger
Air	Gases	Touch	Skin esp. hands	Manipulation	Hands	Planning	Procure-ment	Greed & grasping
Akash	Vacuum	Hearing	Ears	Speech	Throat	Executive capability	Discrim-ination; identity	Ego & pride

TATTWAS AND THE LOWER SPECIES

What then is the relevance of all of this to our theme of the natural world? The answer is as follows.

As we have seen, the five outward physical states are reflections of the same five subtle physical tattwas within our human constitution and it is within the energy domain of these five subtle tattwas that our mental, emotional, psychological and instinctive life has its existence as real energy fields.

It is the degree to which these subtle tattwas are integrated into the subtle mental fabric surrounding the inner life or consciousness of a creature, which determines its degree of intelligence or its level of consciousness. When that subtle energetic matrix is projected outwards as the body of the creature, we then perceive it as a living organism, complete with sense organs, behavioural patterns, instincts and everything else which makes a particular creature so individual.

In other words, what demarcates a species is the subtle energy matrix which makes up the inner nature – and thence the outward form – of that species, whether man or animal, insect or plant. Just as with ourselves, what a creature is, is determined within itself; the outward form is only a projection of more subtle patterns.

So, why is there a difference in intelligence between the species? Why is man more intelligent than apes, and apes cleverer than rats? Why can rats out-perform birds in laboratory intelligence tests and why are birds able to 'out-think' insects and reptiles? And why do we automatically assume that plants have the lowest level of consciousness, even though tests have shown that they too respond to human thought and emotion?

And why are there particular *classes* of creature? Why are there no plants that can move about on legs, with burrowing, but retractable root systems perhaps, or creatures that are half-insect, half-reptile? Or green-skinned animals which can photosynthesize? Why no reptiles and mammals with feathers and true bird-like wings? Why can't man fly? Why is there no smooth, homogeneous curve of creature after creature, each imperceptibly blending into the other in outward appearance and inward consciousness and intelligence? Why are there *individual* species at all and no smooth curve of life forms, the one barely distinguishable from its nearest 'relative'. This would surely be the logical outcome of a Darwinian style of evolution, based upon chance or random mutations.

The answer, say centuries of mystics, is because just as the outer world consists of the five states of matter which have a tendency to separate from each other, creating divisions or surfaces, so too does the inward and more subtle morphogenic fabric that makes a species what it is, consist of these same five tattwas in subtle form. So although these states of energy are intermingled in life forms, all the same, their arrangement into complex configurations still results in the definitive outward patterns that we call species.

Plants, for example, have only the water tattwa active in their inner patterning. Insects are of fire and air; birds of fire, air and water; higher animals of fire, air, water and earth; while man alone has all five tattwas active within him. Note that the word used is *active* tattwas. This classification is discussed in greater detail in *Natural Creation . . . Or Natural Selection?*

Naturally, this is reflected to some extent in the outer body, in the tattvic states which are drawn into the gross physical body of the creature. Plants, for example, often consist of about ninety per cent water.

It is the exact arrangement and patterning of these subtle tattwas, according to the karma of the individual soul, and within the context of the total pattern and rhythm of nature, which determines the body of a particular species. Instinct, of course, is a pre-programmed part of this inward, subtle matrix which makes a cow a cow, or a spider a spider, in both outward form *and* behavioural patterns. A creature does not need to *learn to be* a cow or a spider – or indeed a man. The basic patterns, both inward and outward, are those with which a creature is born.

What about intelligence, then? You will observe that each of the five tattwas is related to the next higher tattwa, with akash holding the prime position as the creative space or vacuum, the matrix of subtle vibration out of which the other four tattwas are derived. Only man has akash. It is the linkage of the inward, subtle form of akash into our human mental energetic structure that gives us the level of intelligence with which we, as humans, are familiar. It gives us foresight; it provides us with the mental capacity to take a overview of things, both in our sensory impressions (i.e. an awareness of space), as well as in our minds. We are able to assess possibilities, to work them through in our minds, considering the potential outcome and to make appropriate decisions. It therefore gives us the ability to think creatively and to discriminate – to manipulate the other four tattwas, making us feel that we are free – certainly more free than the lower creatures. At least we seem to have a perception of choice, however hemmed in by mental habit and socio-psychological conditioning and circumstances we may be.

Just as space or akash provides for the manifestation of the other four tattwas, so too does the integration of this tattwa into our inner make-up give us the mental space, the complete picture, to see how the universe is put together.

It has been said by anthropologists that all of man's culture and civilization have arisen from just two factors: his ability to manipulate objects in a complex manner, and his language. The differences between man and chimpanzee are said to be more behavioural than anatomical – that is: of the mind. But both of these two faculties arise from one common factor: man's foresight, his ability to assess and discriminate, in his mind, the possible outcome of events or the effect of doing certain things.

All these faculties arise from the integration of akash into our inner mind structure. In the outer world, akash is the spatial or vacuum energy out of which all material substance comes into being. In our mind, it gives us the ability to see the relationships between all things and how events might turn out, given certain circumstances. Akash is the point of integration and unification. It is thus the true unified field of physics, while in our mind it is the factor which permits us to see the unitary wholeness within all things.

The capacity for language arises because we can *mentally* perceive the relationships between things and wish to express or give utterance to it. Similarly, it is the origin of our ability to manipulate our environment, consciously and creatively. It is the linkage of akash into our mental configuration which permits this. Without akash, we are animals.

This is why meditation, over time, gradually increases an individual's ability to discriminate wisely, with foresight, and to see things as a whole. Because the more the mind is concentrated, the less are our thoughts divided, and the more we see the Oneness and integration at the true heart of things. The presence of akash in our inner mental make-up gives us this potential.

Man's distinguishing foresight and discrimination thus arise from his ability to mentally project and conceive the possible outcome of events. Events are only changes in *spatial* patterning spun out over *time*. It means, therefore, that we have a greater grasp of space and time in our mental configuration than other creatures. This is conferred upon us by the integration of akash into our mental make-up. Akash is a real energy field or plane in the energy hierarchy of the greater Mind.

A hawk or pigeon may have a three-dimensional eyesight profoundly more perceptive than our own. But this is not reflected in their mental conception of space. They react *instinctively* to stimuli, with no evidence of their possessing any deep *understanding* of space and time. They cannot *figure* things out. That requires the foresight which only man has. Even apes and dolphins do not possess that faculty to the degree that man does. They do not *discriminate* in the way humans can – mentally assessing the possible outcome of particular choices.

All this arises from the integration of akash. Akash – being the formative 'matrix', 'interface' or space of each region in the inner hierarchy of creation – automatically confers the key to our understanding of space and time. For space and time are the underlying framework within which all change and pattern are manifested. For this reason, akash is also translated sometimes as 'sky', not in the sense of our earthly sky, but as the formative energy field 'above' a particular region, from which everything in that region is created or manifested.

Now, the natural law is: the greater the number of active tattwas, then the greater the consciousness and hence intelligence of the creature – and incidentally – the larger the brain required, relative to body size, to bring into manifestation the greater inward complexity of the creature – for communication, language, memory and so on. In this sense, intelligence is the ability to mentally or inwardly grasp an increasing multiplicity of facets concerning the 'outside' world, and then to relate them to consciousness.

Each of the four tattwas has specific characteristics and it is variation within these that brings about the variation of species within a class; ape from rabbit, for example, rabbit from rat, or bat from whale. For apes, rabbits, bats and whales are all mammals, all four-tattwa'd creatures. Species are different from each other because they are inwardly comprised of different tattvic configurations.

Given, therefore, that species have different tattwas linked into their subtle structure, we can expect to find creatures manifesting qualities, behaviour and sensory perceptions which reflect that inward patterning. And this is indeed the case. In the next few chapters, I attempt to show how this is so. Yet this is a reversal of the normal manner of expression within the life sciences, where it is the form which is usually studied, while the consciousness or life of the creature is largely ignored. Here, I am interested – as indeed is the creature itself – with the life and subtle patterning within, which we observe outwardly as the form.

MAN AND FREE-WILL

It is often said that what makes man different from animals is that he is self-aware and seems to possess free-will. As we have seen, both of these characteristics arise from the integration of akash into our inner mental fabric.

Akash is the formative matrix out of which the lower creation is formed and conversely by which the lower vibration is communicated to the more subtle. It is an energy gateway or crossroads in the hierarchy of creation.

When linked into our mental constitution, it is akash which thus gives us a sense of our place within the total tapestry of life. That is: it gives us our human sense of identity. But with our mind running out away from its centre of consciousness behind the two eyes, we lose awareness of our real place in the scheme of things and develop the human sense of identity in an imbalanced manner. This we call egotism – an illusion created by our descent into unconsciousness.

And with the misconstrued perception concerning the real nature of human identity comes our illusory sense of free-will. It is very clear that our freedom to act and think is hemmed in on all sides by physical circumstances and by the habituated, psychological and subconscious content of our mind and emotions. To think that we have free-will is as illusory as the imbalanced sense of identity, the ego, which thinks it has such 'freedom'. The sense of I-ness and the illusion of free-will thus arise together.

If we were to leave aside the question of free-will and examine the nature of the 'I' who thinks it has this free-will, we would find that the paradox of free-will begins to be dissolved, rather than resolved.

As the mind refocuses back at the eye centre, the illusory sense of I-ness gives way to a clear perception of our real place within the scheme of things. Then we are aware of the conditioned nature of our free-will and develop a totally different inner experience of the nature of our I-ness. Then we become increasingly identified with the Divine Source within us. Ultimately, all sense of the little 'I' disappears as the greater self of the Supreme takes its place. It is a question of expanding our consciousness to become the Universal Consciousness. Only then are we truly 'free', one with the Supreme Creative Power within everything.

EGO, THOUGHT AND THE SCATTERED MIND

Every thought we have contains an element of I-ness, of ego. When we sit in meditation and attempt consciously to still these thoughts, then we find how little control we have of our mind, and hence of our sense of ego or identity. A still mind begins to lose its egocentric awareness because it is the agitated mind, full of subconscious mental activity – of I-related thought, motivation and emotions – which automatically generates the mind's sense of ego. The two go hand in hand. Ego implies that the mind is dwelling below the eye centre in a subconscious and scattered manner.

As the mind becomes still, at the eye centre, the discriminatory quality bestowed by akash becomes more consciously manifested and we gain a greater understanding and conscious perception of our place in the scheme of things. This is balanced human identity, a natural part of human mind structure. When the stillness of this balanced perception is disturbed by the mind moving out (e-moting) from its natural centre in uncontrolled thought and emotion, the outcome is automatically a disturbance to this natural perception of things. This distorted perception becomes our experience of human ego. And the greater the agitation, the greater is the degree of subconsciousness, of movement away from the centre, and the greater is the illusory distortion to our sense of being. Hence the expressions 'out to lunch' or 'on an ego trip' and so on.

The solution to all of this is therefore not a psychological analysis of our egocentric personality, for this is no more than ego obsessed with ego. The solution lies in stilling the mind at the eye centre. The imbalanced ego then subsides, like waves after a storm, leaving serenity, stillness and a heightened consciousness.

MICROCOSMIC CONFIGURATIONS – WHICH OF OUR FELLOW CREATURES SHOULD WE EAT?

From the mystic point of view, vegetarianism is concerned with far more than providing a healthier diet. Nor is it simply a matter of sentimentality, though no doubt sentiment quite rightly enters into it: we should have a soft and compassionate heart for the suffering of all life forms – human or otherwise. But vegetarianism is a most fundamental matter, concerning the very nature of life and consciousness itself.

Although the Life Force is the great power behind and within this world, man's reluctance to acknowledge its existence even in himself bears witness to the fact that he has only a tenuous awareness of his own consciousness. The purpose of all meditation and all truly spiritual paths is to increase the awareness of this Life Force within us, to expand our consciousness. To expand it so much that we become one with the creative Ocean of Consciousness, whence everything flows. In the process, we become aware that we have a life other than that of our body. We also realize that it is the activities and habits of our mind which keep us away from deeper inward realization.

As this perception of our own inward nature increases, we begin to realize that the same process goes on in other creatures. We really begin to notice that they, too, are alive or conscious. The more we return to our true inward centre, the more whole and complete becomes our perception of life. It also becomes very simple, because ultimately, the mystic experience of how things are is very, very simple. Complexity lies in the machinations of an intellectual mind and a mind absorbed in the diversity of sense experience and resultant action. The mind becomes intellectual and active because it does *not* understand. When the mind is calm, then intuition develops into constant and permanent, direct perception. That is – the simple experience of being alive and really knowing it, to its fullest extent.

So we become intensely aware of the life and consciousness within all creatures. We also become aware of their mind patterns and processes, and of their sufferings and motivations. And we begin to perceive an inkling of how other living organisms – human or otherwise – experience the world.

The practical problem then arises that for our continued existence in this physical world, we must take the life of these creatures. To find a way of living entirely off the Life Force within, as yogis have done, or as man has probably done in past ages, might seem to be the ideal solution, but this requires a superconscious awareness of the inner process of creation. Something denied to the majority of us. So – though it seems almost bizarre when one really considers it – we have to eat the bodies of other creatures. The question is: which creatures?

The clear and kindest answer must be: those where the least suffering is caused. Suffering goes with mind structure and consciousness. The less complete the tattvic configuration, the mind structure, then the less the potential consciousness of the creature.

This, in fact, provides us with a definition of consciousness as we perceive it at this physical level. For if all creatures are drops of the Divine Ocean of Consciousness, then why do they not all have the same degree of consciousness? Actually, the consciousness, the inward drop of the Lord, *is* the same. But within the Mind realms, in order to contact the 'outer' creation, the universes of the Mind – which include this physical world – it is necessary to have a mind and body structure which contains points of resonance or energetic similarity or contact with that part of the universe with which we wish to be in communion.

This is the function of the tattvic configuration, the *microcosm* of mind and body structure. This microcosm is the mind patterning around a soul which permits it greater or lesser awareness of the greater empyrean. The more complete the microcosm, then the more aware or conscious of the universe a creature will be. And the *configuration* of that microcosm also determines just what the encumbered soul perceives of the universe.

So different species are endowed, in fact come into being, because of different *tattvic microcosmic configurations*. This is what a creature actually is. This is what makes one species different from another. Man has a complete microcosmic configuration and therefore has the potential for complete understanding and consciousness, to realize his full potential as a child of God. Other species are also children of God, but they do not possess the necessary mind-body or microcosmic apparatus to realize it.

All the same, when they are killed, such creatures suffer and their bodies become impregnated with the vibrations of fear, torment and death which those who eat them also take into their own bodies and minds. This is why there should be a sense of understanding sacrament – not wanton destruction – when we take life for our own food.

Whether we take the life ourself or are involved in the chain of demand and supply, we are responsible for the death of those creatures. The killing will leave its impression upon our mind as soon as we become involved, even if only by eating the bodies of creatures killed by others.

The killing, the direct or indirect involvement in their suffering, the vibration of the food itself, all these are etched into the subtle fabric of our own mind. Then that unconsciously formed, mental impression grows like a callus upon our consciousness and we become less and less aware of the Life Force within us, less and less human, even. To be callous is to be unfeeling, to be unaware; we even use the word in our language to mean both the physical and mental coarsening. Whether we kill animals directly or are implicated in the process by buying them, already dead, from the shops, our mind is affected. It coarsens the mind. It hardens it to the finer perceptions of consciousness.

Looked at from a mystic point of view, it sometimes seems so bizarre that one can go out into the human market place and exchange money for the dead bodies of other creatures, though one never finds such wares labelled as 'dead pig', 'dead sheep,' or 'dead hen'. man, at this present time, has become truly unfeeling and unconscious.

We even find people rearing animals in factories as if they were some inert commodity. They really think of them in that way: 'How can we get the most piglets out of this female pig?' Or, 'How can we make this hen lay a maximum number of eggs?' The sufferings imposed upon such hapless creatures are truly terrible. But one can understand how the unfeeling perpetrators of such animal concentration camps can state that the animals themselves have no feeling. For again, man projects his own unconscious mind on to the world about him and asserts to himself and all comers that that is how it really is. While the unfeeling man insists that animals have no feelings, the compassionate person knows that they do. Thus, when we kill, we increase the load of karma upon our own mind, setting the scene for suffering in future lives.

So we have to decide, therefore, which creatures we should eat and we make this decision by considering the subtle tattvic constitution of all creatures. And the lowest in this scale of tattvic and microcosmic configuration are members of the plant kingdom, possessing just one active tattwa – that of water. But yet, the plants and trees are still alive. They communicate with each other and with other creatures, too, as we will see. So even *their* lives should be taken with understanding.

Thus an important result of taking the lives of creatures only from within the plant kingdom is that the mind within us is the least affected. We accumulate the least karmic burden upon our mind. Our karmic entanglement is minimized.

Therefore, if we aspire to an increase in our own consciousness, our own awareness of the Life Force within us, it seems reasonable to look up and honour that Life Force in all other creatures, too. Can we expect release from suffering ourselves when we are causing it in others? By natural law, the two are mutually incompatible. So if we hope to make real spiritual progress, we should satisfy our hunger only from the creatures of the plant kingdom.

Incidentally, it is not being suggested that everyone should become vegetarian. Clearly, the world does not work that way. A principle in nature is being described, that is all. How each individual relates to that is a matter of purely personal concern. And such changes in our conduct of life take place automatically, as we go forward.

STUDYING MIND ENERGY CONFIGURATIONS

I well realize that people of our present Western culture, especially scientists, look upon an analysis of substance according to its physical state as a retrograde step. The theory of the elements is treated as an archaic medieval cosmology dating from classical times. It is certainly true that our modern analysis of substance has gone far deeper than an observation of these outward characteristics. But matter still appears in these five states, just as it always did. nature has not changed. The elements, as states, certainly *do* exist. We experience them all the time.

In our modern scientific analysis of matter, we had to start somewhere. One of the earliest discoveries was that of its atomic and molecular constitution. This gave us some idea as to *how* the elemental states actually arise. It is the *motion* and *configuration* of these molecules and atoms which give us the solid, liquid and gaseous states. And the 'fiery' state is directly related to energetic movement among the molecules and atoms. This was a great advance in observation, worthy of posthumous Nobel prizes, were such awarded for such dated discoveries!

But similarly, we are only just beginning our analysis of mind and of the Formative Mind which patterns all the energies of our experience – subtle or gross.

The five elements or tattwas *are* fundamental conditions or energy states at every level in the Mind worlds. We experience them in their physical condition as solids, liquids and so on. In the mental world of more subtle energies they express themselves with other characteristics: fire as drive, air as the mental idea of acquisition, underlying manipulation and dexterity – and so on. Fundamentally, the tattwas are energies of the mind.

Now, the multitude of creatures which inhabit or have inhabited this planet consist of consciousness, a mind configuration and a body – body, mind and soul, just like ourselves. And just as the body consists of the five elemental states of matter or energy at the gross physical level, so also does the mind configuration consist of the five elements or tattwas in a more subtle manifestation.

At present, we know little more than that concerning how these subtle energies are patterned or configured. Yet there must be a multitude of ways in which they can be put together. For not only are we individual humans so different from each other, but there are millions of creatures all manifestly advertising the differences of their subtle mental blueprints. But the details of *how* these subtle fields of energy are put together are practically unknown to us! This is identical to our situation several centuries ago when our Western physical science was based upon a study of the elements.

But we have to start somewhere. We are as ignorant today of the manner by which the subtle energy fields are put together as we were of subatomic configurations in the Middle Ages. It was just such an analysis of the subtle configuration or anatomy of man which I attempted in *The Web of Life*. many people have commented on the great detail packed into that book, but in fact it barely scratches the surface of the subject.

If we can accept that the five tattwas exist at a subtle level as a starting point in our study of mind configurations as energy structures, then maybe we can move on from there. We can, for instance, see that plants have no fire because they do not get up and go. We can see that man has akash, giving him the potential perception of an integrated world and all else that makes us human. This is just a starting point.

Maybe people will come along who will make a more detailed analysis both from the point of view of subtle energy physics, subtle biology and physiological function, as well as from a behavioural and psychological standpoint. Both approaches are required and should always be integrated.

But to write off the idea of the five tattwas as archaic, purely out of prejudice and unfamiliarity, would be like denying the existence of solid, liquids, gases, heat and vacuum or space itself. If Mind *is* a dynamically active and integrated energetic tapestry, linked in to physical substance, then there has to be a science by which its patterns and configurations can be understood or at least discussed. This, perhaps, is a subtle science of the future, upon which we are only just embarking.

DROPLETS OF BEING IN AN OCEAN OF EXISTENCE

In summary, then, one can say that a living creature is a *being*. The inward essence of all beings is from the same Source. We call it the soul or the spirit. And that soul is a drop of the Ocean of Being, which we call God, or refer to by so many other names.

All creatures are simply droplets of being, swimming and immersed in an Ocean of Existence. The primary reality is an existential one, an ontological one. many physicists have worked their way to this conclusion: that in the end, the solution to the ultimate question must be ontological in nature. This is the essence of true mysticism – for mystic experience is a supreme experience of being and inward life.

We are all droplets of being; we are all microcosms of being – of mind and consciousness – in the macrocosm of Existence. Material science is only an analysis of our sensory experience, while the experience of that analysis – in our intellect, thought, intuition and so on – are themselves experiences outside the realm of sensory experience. But it is *all* of an ontological nature – it is all a matter of *being*.

And the microcosms vary in their inner make-up or configuration. So the microcosm of being that we call a bee experiences life (or being) differently from the microcosm that we call a cat, a dolphin, a kangaroo or a man. The *essential* differences lie in the microcosmic 'structure', in the construction and dynamic processes of the inner being of each creature, each droplet of being.

This being so, we cannot possibly understand the dynamics of body function and form without understanding the subtle 'laws' of being – the 'laws' of mind and consciousness. For it is the inner being which patterns or manifests as the outer form.

3. OTHER THAN HUMAN

THE INWARD LIFE OF PLANTS

Plants, then, are life forms of the water tattwa. Their subtle energetic matrix, their inward patterning, their instinct, is that of water. In mental structure, water provides flowingness and sensitivity. In humans, it is the surging emotionality of feeling. In its yin aspect, it is the soft receptivity of femininity, of true womanhood. This is why the feminine nature is generally more intuitive and sensitive. Astrologers tell us that the water signs are non-confrontational. They flow around events rather than meeting them head-on. It may sometimes make them seem indecisive and easily led, but it also bestows qualities of sensitivity, sympathy and empathy.

Being of this nature, plants therefore automatically respond, to some degree, to the emotional and inner qualities of other species. This we see quite clearly in the example of 'green fingers', where plants respond to the intent and good feeling in the mind of the human caring for them. This does not mean that plants can read our minds, just that they are sensitive to mood and general mental-emotional vibration. Some folk can make plants thrive in the most unusual circumstances, while with others, plants remain static and unresponsive to the processes of growth and life. The same, of course, applies to all interactions between living creatures, including human to human.

This principle is also the explanation underlying the famous experiments of Cleve Backster,[1] when he discovered that his house plants responded by changes in their electrical activity not only to actual physical abuse, but also to his mental *intention* to abuse them. And that they registered similar changes when other creatures – insects, spiders and so on – suffered pain in their presence.

[1]See: *Evidence of primary perception in plant life*, International Journal of Parapsychology 10, 1968.

Emphasizing the power of the inner connectedness of all life forms, especially of those who are attached to each other (as Backster and his plants clearly were), Backster noted how the plants registered the emotional ups and downs of his day in their electrical activity *whether or not he was physically present with them*. For he recorded it on his polygraph lie-detector and compared the timings afterwards. But clearly the plants recorded only *Backster's* emotional ups and downs – not those of the entire planetary population!

Plants have little specific focus for mind energy, such as we see it in the brain and central nervous system of insects and higher species. If plants do have a focus for the Life Force, it lies in their roots, for many plants regularly shed some or all of their above ground growth with the cycling of the seasons, regenerating in the next year from their roots. But many plants, and trees too, can be propagated from stem cuttings, demonstrating the diffusion of the Life Force within them.

All the same, plants certainly act as one integrated life form. The leaves, roots, stems, flowers and fruiting parts act as one and work together just like the bodies of humans and other species. With their 'heads' buried beneath the soil, thrust into rock crevices, seeking the narrow places in tree bark, or anchored in the muds, sands and stones of countless lakes, streams and oceans, plant life is essentially quiescent. The mind is almost dormant, only the watery quality remains connected to consciousness.

Outwardly, this gives them a chemical perception of 'taste' – roots seek minerals and water, while the leaves orientate themselves to derive maximum energy from the sun, 'stored' in chemical bonds. Even their dry seeds await the softening activation of water, sometimes for centuries, before they break forth in growth.

But inwardly they surely feel. Their presence is even soothing to the scattered and agitated minds of men, who create beautiful gardens as abodes of peace, a refuge from the strains and stresses of life. man takes to the jungles, deserts and wild places in search of inward calm and spiritual meditation. But plants do not think, as we humans do; they are not inwardly connected to akash. They have no fire, either, so they cannot 'see' the electromagnetic spectrum like other creatures do. Neither, therefore, can they get up and move about. Their only movement is that of growth, though this is of the most intricate detail and form, a reflection of each individual species' subtle pattern matrix.

To be a plant must be to experience a hazy, quiescent state of consciousness, with focus diffused and little or no opportunity to create new karma. It is a place of restoration for a soul weary from low living as a victim of the mind's continuous pulls and indiscretions of previous incarnations. Or perhaps some plants will always be plants, part of the building blocks of creation. Plant perception is largely of a chemical nature, a diffuse comprehension and awareness of a chemical reality, combined with an inward sensitivity to the subtle watery tattvic vibration.

Here only watery feelings surround the encapsulated soul. From their water is derived their sex, nature's way of ensuring variation within a species through the inner patterns encoded within the nucleus of each cell. For variation and adaptability are essential to survival in a changing planetary environment.

Some folk actually ask the question, 'Are plants alive?' One might equally make the query of many humans! But life is life. The inward Life Force, surrounded by innumerable sheaths and finally by a patterning of the subtle physical tattwas, creates a form or body of the same essential patterning, which is most wonderfully intricate, dynamic and complex, far more than we observe in a stone, in water or in the relatively simple molecular structure of the air.

The German scientist, Fritz Albert Popp, commented at an Oxford seminar in 1989, that he estimated his cucumber seedlings to carry out about *100,000 inter-molecular reactions per cell per second*. An incredible degree of activity. And the levels of activity *within* their molecules and atoms themselves is even greater. Yet it is all ordered and organized in the most exquisite manner.

Compared with inert substance, matter comprising the bodies of living creatures is in a super-dynamic state – highly charged, energetically, taut like a tight string, rather than a floppy one, ready for action and reaction. Such a condition also makes living organisms highly sensitive to their environment, reflecting the characteristics of mind and consciousness themselves.

The building of such complexity requires great energy and so, when the life or soul departs, nothing is wasted and the remaining body becomes the food for other creatures. Thus the natural economy rolls on. Only plants (and some bacteria) can live off minerals and simple chemicals. And they put back nutrients into the soil, too, both directly and by the decomposition of their dead remains through the agency of creatures such as worms, bacteria, fungi, moulds and so on. Without plants we cannot exist, and without the other species, plants cannot exist either. The necessary minerals in the soil are recycled from the dead bodies of other creatures or of plants themselves. Insects, birds and animals are all implicated in the nutritional cycles, the propagative processes and the sex life of plants.

Insects pollinate them, birds and animals carry their seeds in their guts and on their fur and feathers. Plants consume carbon dioxide, releasing the oxygen that we other species require, completing in our turn the full circle by 'burning' carbon and breathing out carbon dioxide.

Even in our own lifetime, within just a few years, the destruction of the Earth's natural forests, as well as man's industry, is increasing the amount of atmospheric carbon dioxide, so subtle is the balance. How, I wonder, do conventional evolutionists imagine that their primaeval world of pure plant life existed in the absence of all other species? They say it existed as such for millions of years. The theory is, of course, absurd, for it takes no account of life itself, the inward jewel of consciousness, around which all forms have arisen, the inner essence that provides the meaning and the understanding. This is His Creation, His Life, His Being that we are all experiencing. He is within us and within all forms, the ultimate pattern-maker within every atom, the essence of all life.

His show is continuous, His presence without exception. He is both far away, yet 'nearer than breathing, closer than hands and feet'. Only the ego has to die, for this knowledge to begin to shine through.

SUBTLE LEVELS OF PERCEPTION

Plants, then, are alive – sentient – they perceive and respond to the environment. And that perception and response has both physical and subtle aspects. As humans, it seems from experience that we can actually perceive at both these two levels – at the gross physical level, through our outer organs of perception, and also at the subtle, more inward, level. The two are clearly allied, however, since we can perceive directly through the mental faculty of sight – subtle sight, one could say – as well as via input through the physical eyes. It is this faculty which sees the subtle physical aspects of the aura, for instance, while in full outward manifestation this faculty is expressed as our physical eyes, which are sensitive to the small section of the electromagnetic spectrum that we call light. Light, of course, is an expression of our *experience* of the electromagnetic energy field itself. Nowhere inside our brains or eyes has any neuroscientist ever found anything remotely resembling our constant everyday experience of light.

Among humans, at this present time in our history, we have largely lost our ability to perceive directly the subtle aspects of our physical world. Due to the overcrowding of thoughts, emotions and subconscious patternings, our attention is drawn down from its princely abode between and above the eyes and plays among the tattwas, gross and subtle, of the physical realm. We are still continuously *affected* by the subtle aspects of life, for they are a part of our inward being, but we are often *unconscious* of this subtle interplay. As children, we are not taught to use our innate perceptive abilities. Rather, the purer observation and clarity of the child becomes conditioned by the attitudes and belief systems of the parents, teachers and associates. The precious gift of consciousness and subtle perception is thus overlain by distrust, doubt, fear and the entire spectrum of human weakness.

The child is even taught, unconsciously, to disbelieve the evidence of his natural peceptions until, as an adult, he fails to notice the emotional and mental life even of his own close family. He thus grows up, largely unaware of his own emotions and thought processes.

'Don't be upset, Mummy,' says the kind heart of the child. 'I'm not upset,' says the mother, thereby throwing the mind of the child into a confusion in which the choices are either to disbelieve the mother or the evidence of the child's own perceptions. So dependence on the mother leads the child actively to believe in what is essentially a dishonesty on her part. Similarly, the lack of clear perception among his or her parents, teachers and fellow children leads the child to imitate their behaviour and inward way of being. The child thus learns to distrust his or her own perceptions – both sensory and inward – moving further and further into emotional isolation, playing out games of egotism and conventional social ritual as a substitute for real life, consciousness and direct perception. In fact, the effect of many past lives upon the child's mind may already be too great a burden for him or her to have ever possessed much capacity for direct mental perception.

Animals, birds, insects, plants and all lower species do not possess man's akashic linkage in their inner equipment or structure. They do not carry the double-edged sword of apparent free choice and intellectual ability, to understand or cogitate upon their place within the scheme of things. Their minds are thus dulled or dormant and suffer a less varied menu of feeling than ourselves. Compensation, however, is present, for it seems very clear to observation that the subtle side of their awareness is not so blocked, if at all, as in their human counterparts.

This is the 'sixth sense' possessed not only by dogs, horses, dolphins and elephants, but by all creatures, even the plant kingdom, as Backster and others have discovered. It is true that many of these creatures are aware of ordinary physical sensations beyond our human thresholds. But within the fields of the subtle tattwas comprising the inner energetic format, matrix or tapestry which makes each creature what it is, each one also possesses subtle perceptive abilities, just as we do. But such perceptions are limited by the tattwas which are active within their make-up.

This is the secret behind many of the stories about animals sensing the presence of man and other creatures, or of rushing to the help of an injured person discovered by means outside the scope of their more obvious perceptive processes. This inward presence of life draws all creatures together into nature's whole. Every creature has this subtle awareness within the ebb and flow of the natural economy. A herd of deer on the African plains may feed undisturbed when a lioness walks among them, if they perceive that she is only out for a stroll. Yet when her carriage indicates that she is hunting, one sight or scent will panic the herd into instant flight.

Behaviourists say that the deer can see this in her body language – in her bearing and the way she moves. But that is only an expression of how the lioness is feeling within herself. With no image of hunting and killing in her 'mind', with no hunger driving her, she does not express such intentions in her actions and demeanour – and the deer pick it up instinctively at both subtle or vibrational as well as gross physically observable levels. After all, the body language of the lioness is interpreted within the mind of the deer.

All the great hunters and stalkers – whether of white or brown skin – have had the ability to consciously or unconsciously 'close down' or quieten their mind and body, as well as to conceal their activity while stalking or awaiting their prey. This is the hidden side of knowing how to move quietly, an art known to all human communities who live close to nature. Silent steps but a noisy mind are not true quietness. Frequently, too, such people have been blessed with the 'sixth sense', their quiet mind automatically transmitting the hungry intentions of a hidden tiger[2],

2 *The man-Eaters of Kumaon* and *The Man-Eating Leopard of Rudraprayag* by Jim Corbett contain many such instances. See Chapter 7.

for example, for all creatures are linked at the level of the mind, as well as soul. We should learn to trust our intuitions, for this is a part of our human heritage. The stiller the mind, the clearer our perceptions become. There is no need to go self-consciously looking for intuitions, for then one can be deceived and become the victim of imagination, but such faculties are automatically rediscovered as we learn the art of meditation, of stillness, and true mind control.

Lorne and Lawrence Blair, who lived awhile with the forest-dwelling Punan Dyaks of Borneo, wanted to accompany the Dyaks on one of their hunting expeditions[3]. But the Dyaks only smiled. 'You are too noisy,' they said. 'You must first become *like* the forest, a part of the forest dance.' This means, firstly, in their *minds*. Then the body follows suit. And they added, 'You would need at least four years just to learn how to breathe!'

Perhaps, too, such inner quietness underlies the legends concerning herbs that make a person 'invisible'. Maybe such herbs are natural neural tranquilizers, no doubt possessing a far more harmonious interaction with life processes than those presently produced by drug companies. Indeed, lady's slipper, skullcap, valerian and other herbs have long been known by Western herbalists to have such calming effects.

Nature has a solution for everything. This is axiomatic, for how else is balance maintained over the millennia? Modern man sees life as separated into compartments, as mechanistic interactions, where consciousness has no meaning. But life *is* consciousness and cannot be suppressed for ever. Everything that exists is one whole within the greater life of the Supreme Consciousness. If our tunnel vision sees only disconnected fragments, the fault lies within ourselves and it is in remedying that ignorance to which our main efforts should be individually devoted.

This potential for subtle perception was even demonstrated by Backster's plants. In one of his experiments, lie detectors were connected to two house plants that had been present in a room at the time of a murder. Backster had previously shown that his own house plants were capable of responding to a lie by changes in electrical activity as measured by the lie detector. The staff present in the building at the time of the murder were then questioned individually under the assumption that the plants would register the high level of emotion in any guilty party when he or she was

[3]A film of their expedition was presented on BBC television during 1988.

questioned. In fact, the plants showed no response to any of those questioned and it was later discovered that the murderer was indeed none of them. An inconclusive test, though Backster's prior research is of considerable interest.

But the irony is that a human being, with all his potential capacity for understanding, is actually so cut off from his fellow humans that a plant sometimes has better perceptions at the subtle level than he has! Such perception, and far more besides, is our true human heritage, but in this present age, 'modern' man does not employ it consciously, though many of the more simple tribes, living close to nature, most certainly do. We are so lost in the tangled impressions, thoughts and emotions within our own minds, that we have temporarily lost sight of the subtle.

Actually, plants are not the only creatures that can be better judges of character than many humans, for stories abound of dogs, horses and other creatures instinctively picking up the unpleasant and threatening vibrations from some human and responding with a personally directed snarl or with a well-aimed kick or bite.

Those who 'have a way with animals' are essentially kind of heart and wish them no ill – a mental vibration to which both animals and humans respond with trust.

THE LANGUAGE OF THE LOWER SPECIES

The language and communication of all creatures takes place at both these levels, too – the subtle and the gross. But by language I do not only mean that of spoken words or written symbols. This would appear to be unique to man. I refer to all forms of *communication*. man's conceptual and intellectual ability can become his worst enemy for he even becomes unable to utilize fully his five outer senses, let alone his subtle potential. We pass so much judgement in our minds concerning the events of a day that we fail to see what is right under our noses. We fit our perceptions into pre-arranged patterns and belief systems, subconsciously *interpreting* all that we experience, thereby losing the meaning and the life within the direct experience. We lose its simplicity, its beauty, its joy, its expression of the inner source. Then we fail to see the beauty in a leaf or in the clouds. We miss all the subtle nuances of facial expression and the almost desperate attempts of the natural world to initiate communication with us.

Man communicates by word of mouth and by body language. Our mind and emotions are automatically expressed in the attitude of our body, our facial expression and our gestures. Everything we do and say is reflective of our personality and mental-emotional patterns. Those who are more aware, also become conscious, automatically, of the vibrations or atmosphere of people and places – of the moods, thoughts and emotions of others. As a good friend of mine once commented, 'If man could only see his aura, he would never go to war.'

We communicate our inwardness in great waves, all about us, carrying it with us wherever we go. This is also communication, and it affects others, too, automatically, whether we or they are conscious of it or not. It is therefore incumbent upon all humans to attempt to control our minds, to pursue some path of mind control, meditation, prayer, yoga or whatever – according to one's inclinations. Our quality of mind and emotion is more important than the clothes we wear or the knowledge we possess. If we think that the outer garments are of importance, how much more important, then, is the subtle auric garment with which we surround ourselves and which actually penetrates the minds and subtle structure of others who come our way – human, animal, insect or plant.

Similarly, among the lower species, their communication is by sound, by smell, by sight, by a multitude of strange senses unfamiliar to our human minds, and by gesture and by inward intent. All of their senses are involved. The alarm calls, the social noises, the sexual or territorial 'come-hithers' or 'get thee hence' gestures – the whole spectrum of movement – is all a part of their language, a continuous expression of the inward patterns within the subtle tattvic mind configurations comprising each species. Together with the vibrational communication at the subtle level, all these are going on continuously before our eyes, if we had the understanding to perceive it.

The tiger and the sambur may drink together at the same pool, for there is a sanctity observed by many animals at the jungle water hole. Or those voracious fish-eaters, otters, have been seen to play in the same pool as the salmon, fish who would normally flee from the presence of such a competent and fearsome predator. Clearly, they know instinctively when the otter is actually hunting. It is expressed in the otter's body language. But why does the otter not take advantage of this sense of security to launch an attack?

Nature is balanced. No creature is given such complete supremacy as to dominate all others, even man. 'If all the beasts were gone, men would die from great loneliness of spirit,' wrote that great American

Indian, Chief Sealth, in his famous letter to the American president, Franklin Pierce.

Creatures have no duplicity, like man. Their intentions are always clear and straightforward. Even the trap-door spider with his hidden and camouflaged lair is straightforward and 'honest' in his blood-thirsty intentions. As indeed is the lion, the otter or the domestic cat when they are out a-hunting.

Each creature understands his own kind the best. Even man understands his fellow humans better than the language of the animals and other creatures. Our senses and subtle mind structures are different. The complex touch and pheromone-based social life of the ant is a mystery to the bee. They have different instincts, different pheromones, different patterns of behaviour, a different mind structure. They do not interact so very much. They perceive things according to their own inward nature and outer form.

We may observe the scenting and postural greetings of dogs with whom we live every day. We may even analyse such behaviour in semi-anthropomorphic terms. But we cannot smell what the dog can smell, nor interpret such scents, body postures and barks in the way our dog does. We do not have a mind which is formed like theirs, we do not experience things in the way they do. The language and the inner mind are part of a whole that makes a creature what it is. We would understand the bark of a dog, like we do our own language, only if we had the mind of a dog. Language is only an utterance, an 'outer-ance' of what lies within the mind of a creature.

Species, of course, do communicate with other species. There may even be communicative and behavioural symbiotic links built in. Even man and dog can become a close team, working together most effectively. Yet the full range of communion available to fellow members of a species, even between strangers from different parts of the world, does not seem to be normally available across species boundaries.

Our physical bodies and subtle structures separate us, one from another. Actually, as long as the soul is associated with the mind, we are separated from each other. But we identify with this physical separation even at a mental and spiritual level, where such separation has begun to dissolve. No two humans ever communicated to the fullest extent if they did not empathize with the mind of the other. And no psychological analysis can ever create this subjective sense of empathy. It may well obscure it.

Similarly with all other creatures of this world. We will never understand them purely by an analysis of their physical form and behaviour. Yet with a quiet, observant and empathetic mind we may begin to see what is before our eyes, before our mind, actually. But no university degree, or years of analytical research will of themselves open to us this magic door of perception. This is a gift to a loving and understanding heart. And to a heart that has earned it, too.

LISTENING WITH ONE'S BEING

Laurens van der Post thinks along the same lines. In *A Walk with a White Bushman,* he is in conversation with Jean-Marc Pottiez, talking about communication. He comments:

What moves me very deeply about primitive peoples is that they still attach an enormous importance to a certain kind of communication which we have lost; and that is that they allow the being of the person they are with to communicate with more than words. They seem to let the soul of the other person – or the animal – communicate by the way it expresses itself, in the look and in the bearing, in the tone and in the voice. They allow that part of communication to play an enormous role; whereas we in the West tend to let words play an excessive role. We forget to listen to the tone and the expression which are used, and these are vital because we tend to use words in a fraudulent manner.

Words demand to be treated with great truth. We feel it is such a terrible crime in the West to conterfeit money; but it should be a worse crime to counterfeit words. We see it in the newspapers and we see it among politicians particularly, who say things they really do not mean, or things they do not even understand. . . .

It is a counterfeit of the spirit that goes on daily, you see, made worse because it has been joined by the visual counterfeit possible in television. But primitive man guards himself against that by listening in to the being of the person who is uttering the words and by adding this to his evaluation of the words.

Some of those primitive men are so good at this that they can exactly imitate the different animals, and even other human beings.

I am thinking particularly of one of the Bushmen you knew who was so good at imitating a professor.

Oh yes, the professor! That was a Bushman! After a while he became more the professor than the professor was himself – just by observing the distinguished man and taking his being into and upon himself . . . yes, I have never forgotten that.

I remember one occasion also when I was tracking an animal, a buffalo, with a Bushman. We were tracking it because we needed food very badly.

The Bushman was following it, but suddenly he swerved aside and he went off in another direction and I followed him, although I could clearly see the buffalo hoofmarks leading off in the opposite way. And then he stopped behind a bush and beckoned me over, and there was an antelope standing ahead, which I shot. And I said to him, 'But how did you know an antelope was there?' He said, 'Suddenly, following the buffalo, my eyes were full of antelope; and I went where the fullness in my eyes came from, and there was the antelope.'

Do you think they have some psychic power unknown to us?

No, I think we all have it in us, but they take intuition much more seriously than we do. I remember an occasion, some years ago, when the front door bell rang here and I opened the door and it was my secretary coming for work in the morning. I looked at her face and she said to me, 'I've got something to tell you.' And I said to her, 'I know what you are going to say – you're going to tell me that the mother of X committed suicide.' I had not been thinking about him, I had not even seen him for months. And she said, 'How did you know?' But it just fell into me as I stood there with her. I do not regard myself as clairvoyant, but I think it shows that there are other ways of communicating that are important. This would not seem strange to primitive people at all.

THE ECONOMY OF NATURE

Creatures see what we call the physical world in so many ways. And the ability of many insects to see the infrared and microwave emanations of particular molecules or of particular plants, for instance, is intriguing in its wider implications as it relates to the overall balance and economy of nature. So let us consider it, as an example.

All matter at temperatures above that of absolute zero emits infrared radiation. This is radiant 'heat'. So it is quite possible that the hairs on plants, indeed their entire shape and form, act as tuned aerials, as vibrational emitters and receivers of highly specific wavelengths, in the infrared region of the electromagnetic spectrum. (For simplicity, we are saying infrared, but actually microwave frequencies are also involved.) Similarly, the gaseous molecules given off by plants may be smelt by us humans, but can be identified by some insects from their characteristic infrared emissions. This we mentioned briefly in the last chapter. Insects who feed off, pollinate or whose lives are otherwise associated with these particular plants have a corresponding detection system in the structure of their antennae, through which they perceive specific patterning in the infrared, just as we perceive what we call 'visible' light, through our eyes. They are not aware of *all* such emissions, however, from *all* plants and other creatures – just those with which they have close ties.

They use the narrowest bandwidth radio frequency, transmission and reception yet invented. And with the millions of insects buzzing, crawling, flying and swimming about, all needing to communicate with their own kind and with other associated species, but not with the millions of other creatures, this would seem to be essential to avoid confusion.

This is a part of the great design apparent in nature. Imagine what it might be like if, as humans, we had not only the senses of other creatures but the ability to understand these communications as they do. We would soon be in input overload and go nuts! Each creature perceives just exactly the right amount of the energetic interactions which comprise the physical universe and its creatures.

Philip Callahan has demonstrated that ears of corn, for example, absorb the moonlight within the humanly visible spectrum, but actively emit it once again at specific frequencies in the infrared. The corn earworm moth, whose hungry larvae can decimate an otherwise healthy crop, has antennae tuned to these as well as other specific frequencies, with the result that when the female flies about on a clear moonlit night, she 'sees' – through her *antennae* – the whole field lit up, like an array of a myriad natural light bulbs.

And perhaps the light from those ears of corn which are in just the right condition for her to lay her eggs, fluoresce in a particular manner, recognizing which from her own inward pre-patterned instincts, she homes in and deposits her eggs.

Now, a Western, Darwinian-style, selfish model of nature – a reflection of man's own attitude to nature – would say that the moth evolved independently of the plant whose leaves it wished its larvae to eat. *But then why didn't the plant evolve so that it no longer attracted that insect?* Surely this would have been the 'logical' outcome if nature were so selfish? But this kind of observation is flawed. For nature, Gaia, is a whole; created as such and working as such. Physical nature is the outward aspect, the outward shell, of the great Egg of the Mind, the one Golden Womb of creation. Integration and wholeness are thus intrinsic.

Seen as a whole, nature is not selfish. There is enough to share around if no species gets out of hand. And if it does, its own activity carries within it the seeds of its own self-balancing. If the food-supply is curtailed by an increase in population, then the population is automatically reduced by that very lack of food. If man upsets his planetary ecosystem, then he is dirtying his own back garden which will automatically recoil upon him and curtail his activities. Selfish motivation is inherently negative and is reflected in selfish ways. And these self-centred ways automatically cause suffering and, ultimately, curtailment of the activity.

If I continually throw rubbish out of my back door, sooner or later I will suffer for that and will have to get out there and clean up the mess, as a matter of survival. The balance of yin and yang are present throughout the universe and in the natural balance of all creatures upon our planet. And man is not separate from nature or immune from the consequences of his own actions.

Nature is not selfish in her wider workings. A plant and its flower may look beautiful to our human eyes, but a plant itself has no eyes with which to perceive itself or its fellows. 'So what use is beauty to the plant itself?' the reductionist and materialistic mind might say. How then, can a plant have *evolved* beauty? But its beauty and its form are a part of the total economy of nature. It lives not for itself alone but as a part of a whole. Insects which may contribute to its survival will see it one way, man sees it in another, a seed-seeking bird will perceive it in another. But which is the *real* plant? What the plant perceives of itself? Or the insect? The bird? Man? It is not a question with an answer, for all these 'realities' are illusions in the wider sense. There seems to be no absolute reality in the world of physical forms where every creature sees things differently.

Similarly, those insects possessing an ability to see things through infrared or heat emissions are in an interesting position, for while we humans perceive airborne molecular vibrations as sound, such a creature would have no need of ears, though many do possess excellent hearing, for it may be able to *see* some airborne vibrations as molecular agitations or patterns.

So while crickets, grasshoppers and cicadas are easily *heard* by such creatures as ourselves, these same airborne vibrations may actually be *seen*, as well as heard, by the fellow members of their own species, much as we are able to see the air shimmering with heat as it rises off hot rocks. And we cannot say that one way of perceiving things is more real than another.

And then, with such apparent integration in nature, how could such a *whole* tapestry have evolved piecemeal? How does 'chance' give rise to order? A mosquito may see the scent of our bodies and be attracted thereby. But in the process of evolution, which came first? The ability to perceive the scent, attraction by that scent, or the ability and desire to drink blood? Of what use would just one of these faculties be without the others? How could a creature so specialized as a mosquito ever evolve? Their senses, instincts, behaviour, anatomy, physiology, digestion, biochemistry – the whole creature – is surpremely designed and integrated for its bloodthirsty life. How could such a creature have ever evolved all these highly integrated facets of its being in purely random steps?

Again, how – if at all – does a mosquito perceive its own high-pitched ping as it closes in on its prey, a sound not uttered as it dances with its fellows in gay abandon in a summer evening's cloud. Since this warning ping frequently results in its untimely demise, or at least in its loss of a meal, one would imagine that, were a naive perception of Darwinism strictly correct, mosquitoes would have long since evolved soundless flight – like owls, for example.

Of what use to the mosquito, one may ask, is its ping? From the point of view of natural creation, however, dare we suggest that a not altogether unfeeling God introduced the sound as a warning to its prey, so that in the natural economy, mosquitoes did not have an unfair advantage and that all warm-blooded creatures in mosquito-infested areas were not permanently condemned to a life of itching and scratching!

Similarly, how does a moth perceive the ultrasonic acoustic radar signals of a bat? Certainly, moths duck and weave in their escape attempts, for which their excellent wide spectrum sight is clearly advantageous. But do they *see* the radar signals, too? And can they – by clever use of wing vibration or other sounds and squeaks, jam the bats' radar by emission of confusing signals?

Indeed, there is a lot going on out there in my garden, which I do not understand.

HIGHWAYS, BYWAYS AND ROAD SIGNS

So all creatures have their own world of physical reality, a world defined entirely by their mind configuration. One sees evidence of this everywhere.

Sitting at breakfast on the verandah of Bombay's Sun'n'Sand hotel, some while after reading Philip Callahan's *Tuning in to Nature*, I found myself quietly observing a highway of tiny, high speed ants as they took a specific path to and from their nest under the edge of the patio, gathering up enticing goodies from further down the flowerbed.

As we have said, it is well established in scientific circles that ants and many other creatures, too, communicate by means of external molecular signals known as pheromones. From Callahan's work, it is clear that some, perhaps many, insects will 'see' these molecular communicators by means of their tuned antennae and the laser-like, infrared molecular emissions. Ants lay down these chemical markers on the road to a food supply so that they and their fellows may know where to go. But it is a probability that they perceive them, not like a dog sniffing its way along a scent path, but as a motorway, illuminated by very particular lighting signals.

So had these cunning little creatures marked out their pathway to food with a species-specific street lamp system that only they could perceive? The 'colour-pattern' of this illumination quite clearly said, 'Food – this way!' In fact, since ants, like other insects, have a variety of these illuminated chemical communicators, each meaning different things, their highways – especially within their nests – must provide them with a variety of 'colour-pattern' coding and meaningful messages.

Watching these high speed ants on the Bombay verandah, I was impressed by the fact that ants travelling in different directions upon their highway stopped dead upon meeting, touched each other and then continued upon their way. Further messages were clearly being transmitted, perhaps confirming the continued existence of the food supply, affirmation of being members of the same colony, and so on. A truly social and self-supportive activity.

Imagine the myriads of such infrared and other pathways and signals that the various insect species must be utilizing. Their specifically tuned antennae receive only what is relevant to their existence. We humans lay out our highways and byways in our fashion and each species does so in theirs, and we mostly remain unaware of each others' very own and personal world of perception. Our homes, our territories and our pathways are all superimposed and coexistent, the one hardly suspecting or aware of the existence of the other.

Every environment is an economy of such coexistence, a multitude of interrelationships and separations, of symbiosis and selfishness. It operates as an indivisible, adaptable whole which finds a dynamically shifting point of equilibrium according to circumstances and the cycles of nature.

All creatures contain, within themselves, their own individual characteristics of mind and being, determining their varying levels of intelligence. They are, quite clearly, highly organized and know what they are about, though it is difficult for us to reach into their world and see it from their point of view. But we are given an insight into the limitations of their cognitive capabilities when we observe a spider responding to a vibrating tuning fork by attempting to wrap it up for food. Or by a robin quite determinedly attacking a bundle of red feathers that it has mistaken as an invader to its territory while ignoring an adjacent and perfectly good-looking stuffed robin, but lacking the all-important red feather flash. Their instincts are a tight binding of mental or subtle structure and they are constrained to act accordingly.

No man, nor even a dog, a monkey or a dolphin, would be fooled quite so readily as the robin, though we too react according to habituated mental patterns. Bluetits may be famous for having figured out how to remove the silver foil caps from our early morning milk bottles, to find a creamy delight beneath, but I remember watching with amusement as one of our local bluetits swooped down upon a friend's Mercedes sports car parked in our driveway and perseveringly attempted to remove the central silver disc from the Mercedes logo on the front bonnet. He came back several times, quite convinced that only a little greater effort would supply him with the treat he had in mind. Like the robin and the bunch of red feathers, our little friend was clearly unable to tell the difference between a milk bottle and a Mercedes! Not the sort of mistake you would expect from a mammal or a man. Clearly there is a difference in the characteristics of the mind possessed by each.

DOGS AND DREAMS

Anyone who has ever kept a dog must have observed its clear indications of a dream life. Whimperings, body movements, tail-wags, even little yelps are all heard to emanate from sleeping canines. This is a clear indication of a 'thought' life independent of bodily activity. 'He's chasing bunnies', we sometimes say, affectionately, not realizing the import of our observation.

The nature of that doggie world of thought is closed to most of us. The degree of empathy and perception is difficult to muster.

But they, like all creatures, clearly have their own inner life. They express a wide range of emotions and feelings to which we can readily relate: *expectantly waiting* for their master or mistress to return home, display of *affectionate greetings*, or *excitement* when it comes time to go out for a walk or prepare their food.

I once knew a Yorkshire terrier where even the word 'walkies' sent him into gyrations of joyous bouncing and tail wagging. This was followed subsequently by great interest in all the smells and scents of the roadside, even the occasional canine encounter – accompanied by the whole world of doggie communication: nose to nose, nose to tail, side by side, tail and body postures, ears up, ears down – their inward mental structure is quite clearly being expressed in a language and through sense perceptions and actions we do not share.

And when at rest, does man's best friend lie in contemplation of his next act, or of the possible menus for his next meal? What passes through the mind of a dog as he snoozes with half an eye upon his trusted human companions – whimpering, tail-wagging and yelping in his sleep?

And why is it that dogs have become one of man's closest animal allies? Why not other animals? It seems to be because, in their doggie way, they share a similar range of social emotions and behaviour and have a subtle structure in sympathy with their human companions. Wild dogs and wolves have a strong social bonding within a pack, and to a domestic dog, we humans are probably just a rather special part of their pack. For in general, we are also social creatures. I do not mean that they do not differentiate between dog and man, only that we have become integrated into their already existing, instinctive mental-behavioural patterns.

Wild cats, for instance, are not particularly social creatures and these tendencies are displayed in the behaviour of our domestic pussies. Dogs are generally more expressive of their feelings than their feline protagonists. Pussies are different creatures, within themselves, more self-contained. They have different mental processes from dogs.

And one must also remember the effect of previous lives. It is from the mind energy patterns, the personal associations, actions, desires and so on, which all lie in the mind, that the next life of a soul is determined. Processes of genetic inheritance are only a part, at the physical level, of a multidimensional formative and creative system at work, but it is the mind which automatically forms the patterns which lead a soul into its new body. Indeed, the mind actually forms that body and the life's destiny, as we will see.

Returning to our canine theme, dogs still carry with them the same instincts possessed by their wild ancestors, still remaining members of that same species – the wild wolf – even after so many generations of breeding have made them what must be the most diverse species upon Earth. The herding, hunting, guarding, group companionship and communication possessed by the wolf and many wild dogs of today are almost identical to those same instincts expressed by our domestic dog.

If the materially minded evolutionists were right and mind patterns whether in humans or animals are genetically transmitted as nothing more than expressions of biochemistry, then why do dogs still retain the same instincts as the wolf? Why, indeed, are they still the same species? If evolution is simply a matter of selective breeding, then why can we never produce anything other than dogs from dogs, however hard we try? There is clearly something more to dogs, and all living creatures, than meets the material eye! And that something is nothing more than their inward consciousness, heavily hemmed in by their subtle and instinctive mental processes.

All who have kept wild animals as pets or got to know them in the wild have come to realize that they each have their own personality and manner of expression, within the constraints of that species' natural instincts. But expression of what? It must be of their inward mental and subtle structure.

The body of a cat when hunting takes on a quite different demeanour to that of the same pussy when out for an idle stroll or with some other intention in mind. Just like humans, unless we think, we do not act. A mental activity, shape or pattern is required for the outer action and behaviour to be manifest. How is it that we have blindly come to think that animals and other creatures have no inner life when their every action tells us that they do?

My father used to train and breed English Springer spaniels as a hobby. They were trained as gun-dogs unfortunately, but are nonetheless a very affectionate breed. The trick was always to spot the most intelligent even when they were still puppies. For my father would keep the best for himself, to train and breed, and sell or give away the remainder.

One judged the pups purely on appearance, behaviour and demeanour, as one would one's fellow humans, becoming quite skilled at it as time passed. Sometimes one might also set them trivial tasks requiring a certain degree of intelligence to solve some problem, and the more intelligent ones would thus demonstrate their superior ability. A good sense of smell and an active intelligence were the factors which made a good dog.

Many a time he commented upon the mental rapport he found with his most intelligent animals. His most famous champion, Jasper, would even turn and, by his demeanour and stance, 'tell' my father that he was not right, when directed to go in a wrong direction in pursuit of some unfortunate, injured but running, bird. He would stand and just look at his trusted trainer, not moving off as directed. My father learnt to understand and trust the dog's perceptions. 'Go on then,' he would say, coaxingly, and off Jasper would go in the direction he knew was right.

So it is not only the inward patterning of particular species which makes creatures different, but *individual members* of a species are different, too, and most species can distinguish fellow members of their kind by sight, sound, smell or other senses and probably by behavioural patterns too, just as we do among our fellow humans. We know people by their walk, their gestures and their general body language. Each creature has its own karma, its own personal mind patterns, derived from its own past history of previous lives. These it brings with it to its new life. It is these patterns which are responsible for the body it now has fashioned, as well as for the individual 'personality' of each creature.

INTERLUDE

SEEING WITH THE EYE OF MIND

Through spiritual vision he [the mystic] will see in his mind spiritually all the visible things which are seen by others materially. And inwardly in his mind and in his thoughts, he will survey all the present creation and the worlds that have passed or are still standing; the years of the world with all the happenings that occurred in it, and the men with their wealth and their power; the revelations of the [spiritual] benefits which were bestowed on the Fathers, and the retributory judgements that took place generation after generation [or 'birth after birth'], together with all the various vicissitudes which the affairs of the creation undergo.

All these things which a wise man [of the world] sees materially, he strives to investigate in his mind spiritually, through spiritual vision. He does not see the different plants like an agriculturalist, nor the medicinal roots like a physician, but everything that he sees with his material eyes he secretly contemplates in his mind through spiritual vision.

In this way, the mind is taught and instructed to look inwardly at the spiritual nature [of things], towards the secret power that is hidden in everything and works in everything in an incomprehensible way.

(Simon of Taibutheh, East Syrian mystic, died about A.D. 690.
From: *Early Christian Mystics*, translated by A. Mingana.)

4. STRANGE SENSES AND MULTIPLE REALITIES

MULTIPLE REALITIES, MIND STRUCTURES AND
PATTERN MATCHING

Despite all our scientific study of the physical world, one can still look with wonder at the panorama of 'ordinary' objects and events presented to us by our senses and can still – indeed should – ask the basic question: 'What is it all, fundamentally?' And what, too, am I? Without an answer to that question, any answer that comes from that 'I' comes from a point of ignorance and is thus intrinsically suspect.

All physical manifestation is a dance, an effect, patterns in space giving an appearance of 'reality' and linked to our consciousness through our sensory perception. But, as we have seen, different species have different perceptions. They are linked into this patterning in different ways, according to their own inward subtle or 'mental' structure. There is thus no absolute physical reality. The physical world differs for every species.

So some insects perceive infrared and ultraviolet. Flies taste with their feet, planaria (aquatic flatworms) unerringly follow scents in water, sharks monitor the electrical nervous activity of their prey, knife fish use electrical radar (three-dimensional echo location), electric eels hunt the knife fish by homing in on their radar bleeps, while migrating birds are sensitive to the earth's magnetic field and polarized light. Dogs find the world a place of infinite aromas; bats, whales and dolphins use acoustic sonar. Even plants communicate with each other.

What one creature smells, another may sense as heat, while yet another may see it. Presumably, this is what moths are doing when they fly into a candle and are burnt to death. They see the heat of the candle which looks like something to which they are instinctively attracted – perhaps a food plant, or a flower, or a member of the opposite sex. And too late do they discover that the flame burns and is lethal.

But which is the 'real reality'? What we perceive or what *another species* perceives? Even among humans we all react differently to sensory, social, intellectual and other experiences. So which is 'real', or do none of them possess any ultimate 'reality'?

We cannot say that there is a separate, solid, objective world which different creatures perceive in different ways. Our physical world consists, is entirely made up, of these perceptions. Even mass and solidity are only sense perceptions. And the perceptions are subjective, in our mind. The 'world' to another creature is a quite different place. Yet we cannot claim that our sensory reality is more real than theirs: that we see it the 'right' way and that they see it in some less than real manner.

So we are always drawn back to our *experience* as somehow being involved intimately with what we may thing is an objective experience. Experience is subjective, whether sensory, conceptual or otherwise, and our confusion only arises when we think we are *separate* from what we perceive. In fact, in a gloriously mystical way, we are an integral part of this cosmic energy dance and can only really understand it all by mystic superconsciousness, from within.

The fact that our own personal experience of being alive is different to the experience which all other humans and lower creatures have of us, is a clear and obvious pointer to the mind-nature of this physical domain. For all such experience – sensory, mental or emotional – is essentially subjective, in our minds. The apparently outward world, perceived through our senses, is actually of the mind.

Taking this perception to the lower species, we can begin to glimpse how both we and they are constructed, from within-out. Just like ourselves, the real creature's feeling for itself is subjective, within its own mind structure. What a fellow member of its same species sees of it, is quite different from what the individual feels about itself. It is certainly very different from that which other species perceive of it, for the sense organs of different species are constructed differently and perceive things in different ways, as we have seen.

In fact, it would appear that it is the mind and subtle structure of a creature, which – acting as a blueprint – determines the nature and structure of the physical sense organs themselves.

There is a certain wasp, for example, of which the female is flightless. The male seeks the female, picks her up, and mates with her on the wing, later depositing her where she can lay her eggs.

The male wasp, therefore, clearly knows what shape and colour to look out for. He is born with that instinct, for no mother wasp ever taught him, nor does he possess the capacity to learn something so fundamental to his existence.

But there is an orchid whose shape and colouring exactly mimics that of his female, flightless partner. When the male wasp spies it, he wastes no time in the attempt to mate with what he clearly takes to be a female of his own kind. He is quite insistent in his amorous advances despite the patent lack of response.

Like bees buzzing up and down my window pane, the instinct is too deep for such a creature to see or to reason that he is not getting the expected response. Transparent glass is no more a part of a bee's mind set than mimicking orchids are to those of a wasp. They cannot figure it out, and so go on acting out of their inward instinctive patterning.

And the resemblance of orchid to female wasp is not coincidental, for it is the means by which the orchid spreads its pollen to other members of its species. Without the wasp, the orchid cannot reproduce. The weight of the wasp upon its lower lip brings down the upper lip of the flower containing the anthers, to dust the wasp's back liberally before his eventual, disappointed departure to another flower, carrying the orchid pollen with him. The biological engineering is absolutely perfect.

Leaving aside the question of how such a *perfect* and *whole* symbiotic relationship could have ever *evolved*, Darwinian-style and piecemeal, out of 'random fluctuations in the DNA of wasp and orchid', we are left with the interesting observation that the mind structure, or instinct, of the wasp is organized in such a way that it instinctively seeks this particular shape, for the purposes of fulfilling its urge to mate.

This instinct – along with all the other instinctive and subtle inward facets which make up the creature – then acts as the blueprint for the actual patterning, formation or projection of the physical sense organ and body itself. The 'mind' of the wasp, therefore, seeks the pattern of the female through an appropriately structured sense organ, *but the primary experience of the creature lies in its mind and subtle structure*. It is as if the inner mind structure presents a predetermined pattern to the sense organ, through the brain and sensory/motor nervous system, and only that input which relates to the inward pattern is able to gain entry to the creature's world of subjective experience. In fact, its sense organs and body are only an outward expression of this inward patterning, which we call instinct.

Using computer terminology, sensory perception and motor activity are thus only *pattern-matching exercises* between the inner mind structure and the outward physical body. The outer form and the behaviour of the creature are thus always at one – man does not try to flap his arms and fly, any more than an insect seeks out the sex pheromones of another species as if they were its own. The mind structure – its motives, sense perceptions and instinctive drives – these are all reflected automatically in the outward physical form.

That which constitutes a particular species, therefore, is its inner mind and subtle energy structure, a state which lies beyond the level of DNA. And this is borne out by the fact that although the outward forms of creatures can be bred into a most amazing degree of variety, as we find in the domestic dog, the inner instincts remain very much the same. As we have pointed out, from friendly licks to territorial growls and all other social communications, domestic dogs are only displaying instincts still common to their wild ancestor, the wolf.

This pattern-matching exercise of the mind is, indeed, how we humans already live much of our lives. Emotionally and mentally our mind seeks out the things, people and ideas to which it is already habituated. The mind operates largely by habit. Instinct is simply a habit with which a creature is born. The less the capacity to learn and consider, the greater is the habit instinctive. The child born under the influence of certain cultural, social and religious thinking, automatically develops those mental traits and will normally seek companionship among like-minded individuals.

Within a community, we seek out for our companions, those whose thinking is similar to our own. This is easier for us, for it presents less challenge to the habits of our mental and emotional processes. In this respect, like attracts like. It is, again, a pattern-matching exercise.

In physics and science, we follow the same procedure, often unwittingly. We design and perform an experiment, or make observations, according to a preformed set of ideas or concepts in our mind. These ideas or concepts are called a paradigm. But although they are only thoughts in our minds, by the very design of our experiments we project these patterns onto the 'outside' reality and therefore either fail to see anything else, or are confused when observations do not fit our inward mind patterns and expectations.

I considered this process in some detail in *The Secret of the Creative Vacuum*, for if our basic premises are incorrect, then all findings based upon such foundations will also be suspect. And this can so easily be the case, for such premises are instilled into the scientific mind set at an early age, becoming accepted points of reference, though really they are often nothing more than habits of thought.

So science, too, is conducted mostly as a pattern-matching exercise. And with the mind, even of humans, behaving in a habituated, instinctive fashion, many an insightful perception or suggestion is rejected out of hand by the deeply engrained, yet erroneous, patterns adhered to by the majority. Just as a Christian might reject Buddhism without ever studying it or getting to understand the meaning behind the words. Or vice versa.

So even as humans, what makes us human – our basic inward energetic structure or constitution – is essentially instinctive. We know how to talk, to walk, to hold things, to think even – to be a human being – without having to figure it out. Learning is no doubt present, and the higher the consciousness of the creature, the broader the tattvic tapestry, then the greater is the capacity for learning. But it is all within the basic framework of that creature's subtle and physical structure, within the bounds of instinct.

As humans, we may learn to talk, but the impetus to learn and the capability to acquire human language are instinctively human. If a chimpanzee and a canary are raised in the same house as a child, neither of these two companions ever learns to talk like a human. Neither does the child learn to communicate like a chimpanzee or a canary. All creatures communicate only in the manner to which they are instinctively and inwardly programmed. The canary goes 'tweet' and only 'tweet'. In fact, even a duckling raised by a farmyard hen goes 'quack', rather than 'cluck', maybe occasioning some degree of confusion in the hen house! But a duck always behaves like a duck, whatever his social conditioning, upbringing or training! But nowhere is there a gene, or group of genes, which makes it quack. Once could not genetically engineer a duck to make it cluck.

One could say that the duck goes 'quack' because his vocal organs will only say 'quack', and that these vocal organs are genetically patterned. This is partially true, but the motivations for saying 'quack', the *kind* of quack a duck makes, the social message that is being conveyed – fear, alarm, pain, surprise, contentment, 'Come here', 'Go away', 'Look what I've found' – this all requires a mind behind the larynx to make it express or utter the intent and feeling within the duck's mind, to give the quack the correct inflection and intonation. No one can make a dead duck go 'quack' any more than a dead person can be made to speak. Mind, as well as brain, are required in both instances.

It is a question of *meaning*. Meaning lies in the mind, beyond words – just as one may search for a word to express one's meaning. The meaning is there in the mind before the word is found. Meaning is related to mind structure. The baby never learns to go 'tweet' because his mind structure has no meaning for the noise. Neither is he able, usually, even as an adult human, to distinguish the sounds made by individual birds of the same species – though the birds themselves can do this.

MULTIPLE REALITIES AND SENSORY PERCEPTION

Among whales, dolphins and bats, this pattern-matching process can be very clearly observed. It is even taken a step further. Generally, the mind, through its sensory indriyas, provides the patterns which – mirrored in the input characteristics and design features of both brain and the physical sense organ themselves – automatically selects and sorts the incoming sensory signals into what the mind is expecting. It is a multilevel, yet highly integrated, energy system.

In the acoustic sonar systems of these species, however, the mind first patterns the *vocal organs* to produce highly specific sound signals or patterns. These then go out into the environment, are reflected back in a multitude of different ways according to the objects encountered, are picked up by the most amazingly refined hearing organs, matched against an inner mind structure capable of interpreting this data as a full and complete three-dimensional world and used as a major sensory means of perceiving their watery or aerial world. Not to emit and receive these sounds constantly might be our equivalent of the sun's disappearance at night, of someone switching the light off, or of partially shutting one's eyes, though these creatures are also supremely aware of externally generated sounds.

Our human world is a visible one, to which we then associate sounds, smells, tastes and touching. Probably over 80 per cent of our sensory impressions come through our eyes. We therefore, quite understandably, make the error of assuming that the world is essentially a visual one. But for many other creatures this is not so. In fact, few other mammals are so heavily reliant upon sight as we are, where colour vision would seem to be the exception rather than the rule.

Birds, however, are clearly able to differentiate colours. Laboratory and field tests have been conducted of course, but a simple look at their often gaudy hues, essential to their behavioural repertoire, is enough to tell us that they too, perceive fine gradations of colour. And so too, one imagines, do butterflies and day-flying moths. Certainly they are often far more colourful than their night-flying cousins. But even many of these have some interesting hues, at least to our human sight, though just what other creatures perceive of their own or other species is not so easily understood.

As we have seen, flying insects are often particularly sensitive to the electromagnetic spectrum. Certainly their close association with the coloured patterning and individual shapes of flowers bears witness to this. Moreover, there is not so much to smell when you are flying high up and away from the scents and aromas that cling to the Earth – or even if, like man, your nose stands at a height of five or six feet, rarely bending towards the ground. And excellent sight is most important if your world is in the air or your head is more than a few feet above the ground.

Whether you are perceiving the planet from only a few feet up or from around the lofty peaks of mountains and hillsides, excellent sight (or sonar) is almost a prerequisite: there are hosts of obstacles to circumnavigate and far distances to discern if you travel by air! And you need to be able to find food and companions, too. The air leaves no long-lived scent trails, like the earth.

The relevance of such multiple sensory realities to conventional evolutionary theory has so far remained unconsidered. For if our physical reality is largely linked to the mental and sensory mechanisms we possess, then what was the nature of physical reality *before* life is supposed to have spontaneously emerged – like mice out of soiled linen, as some Victorians thought – from the primaeval muds of ancient oceans? If every creature perceives a different physical reality, which is the 'real' physical reality from which those early acellular creatures are supposed to have emerged? Our sensory impressions of the reality of mud and ocean constitute the *whole* of that reality. It is only for humans that they exist in that way. Does it make sense to suggest that acellular creatures evolved out of a 'reality' which is relative only to a human-style perceptive system of mind and sense organ? Their perception of the physical world is quite different to ours.

Some folk might say that although 'red' is a subjective, mental experience, there is an objective reality which we can measure with our scientific instrumentation. But our instrumentation is again only perceived, as well as conceived, in our own human minds. It is related to *our* mind structure, not that of other creatures.

No doubt, there is some measure of 'objectivity' for humans can actually see how well other creatures are designed for the perception of infrared laser frequencies or acoustic sonar, for example, or for flight, swimming or digging. But yet we cannot say that this notional world of human scientific description is the more fundamental physical reality to which all creatures relate.

Because of the integral parts played by Mind and consciousness, it seems that any purely mechanistic description of the origins of life and the physical universe, is doomed to failure, for it ignores the manner by which the physical creation and physical bodies come into being as expressions of Mind and consciousness. These topics are discussed more fully in the last two chapters, on the *Formative Mind*, and throughout the two associated books in this series, *Natural Creation . . . Or Natural Selection?* and *Natural Creation: The Mystic Harmony*.

Back in the world of multiple sensory realities, vocalization and dynamic, three-dimensional echo-location are the most fully developed sensory systems of some bats and cetaceans. This is useful if you fly by night or live in murky waters. If you have ever watched a bat chase a moth, as both wing and weave through the evening air in amazing aerobatic gyrations, you will know how three-dimensional and accurate is their perception of this world. But to bats – both in their mind structure and associated sense organs and motor responses – the world is a three-dimensional *sonar* world – not a visual world. Their sonar is like a personal sonar 'searchlight'. When it is switched off, part of their world is silent, though they will still hear the sounds of many other creatures, especially – one imagines – those of their own kind, their prey and their predators.

Some moths even possess a bat-confusion kit. When chased by a bat – which they evidently recognize as such – they emit bat-like sonar squeaks, which confuse the bat as to exactly where it (the bat) and the moth really are. Radar jamming was clearly invented many millions of years ago!

So the world of every creature is different to that of all others, sometimes greatly, sometimes by not so very much. Taking yet another example, compare the visual fields of man, rabbits and the sperm whale. When looking directly ahead, man possesses only 180° vision. We cannot see anything behind us, (see Figure 4.1). 120° of this visual field is *binocular* – we see the same things with both eyes. Binocular vision is of great importance, for it is two visual fixes on objects in a three-dimensional world which give us a perception of distance. Try judging distances in an unfamiliar place with just one eye open, for instance, or notice how much more difficult it is to put your finger on a particular spot when one eye is shut.

For about 30° on each side of this 180° field of view, our vision is monocular. It is called our *peripheral vision*. Here, visual acuity is not so sharp, nor is our ability to judge distances so good. But this peripheral vision processes input very fast, with visual information reaching the brain and mind more rapidly than from the binocular area. This is handy, for we can thus spot potential danger and react rapidly to input gleaned out of the corners of our eyes. And turning our head directly to the source of the new input, we can then decide – if we have time – upon appropriate action.

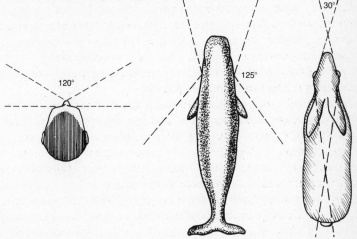

Figure 4.1 The visual fields of humans, sperm whales and rabbits. (After Bonner)

You can check this out for yourself very easily next time you watch television. An image on a TV screen is not continuously present, but is relayed as a series of stills, twenty-five to a second. This is fast enough for our central vision to be fooled into the illusion of a continuously moving picture. But if you look off to one side and view the television with your peripheral vision, out of the corner of your eye, you will be aware of this twenty-five frames per second illusion: you will see the image flickering.

Man, of course, is fascinated with himself and how he perceives and does things, and many interesting experiments have been designed to test out how our perceptual and motor responses function. Many experiments have been performed upon other creatures, too. But unless one realizes that there is a mind structure, including appropriate sensory and motor indriyas at a subtle level, as well as corresponding sense organs and motor functions, such experimental observations remain obscure and lacking in a full understanding and interpretation.

If you are a creature that needs to locate things rapidly in a three-dimensional world, whether as a hunter or for other purposes, binocular vision or 'binocular sonar' are essential. man can, to some extent, tell where sounds are coming from, though we are often fooled by reflected sounds. Trying to locate a particular noise in a built-up area, for instance, can be tricky, since sounds are reflected off buildings and appear to come from places other than where they are. Our three-dimensional appreciation of sound is not so very good.

But with vision it is another matter. We are not so easily fooled by reflections in lakes or puddles. We know them immediately as reflections, just as we appreciate *shadows* cast by objects as they obstruct the passage of light. Indeed, we unknowingly use the variation of light and shadow in our environment to tell us more of its structure and three-dimensionality.

But sound also casts sonic shadows, though other than noticing that shutting the door can reduce the sound, we do not pay much heed to it. Additionally, since sound is reflected differently from surfaces of varying density, a good quality ultrasonic sonar would bestow some degree of 'X-ray' sonic vision, infinitely more refined than our present day ultrasonic medical diagnostic instruments.

But if your main organ of sensory perception were sound, and you had an excellently organized mind structure and sensory system for appreciating three-dimensionality and structure from sonic input, then the world would be fully perceptible by day or night. It would make little difference what time of day it was.

An ordinary shut door, for example, would not be opaque to you, if you were one of those species of bat or whale who relied heavily upon detection of sound waves! Sound would come both through the material of the door itself and around all the edges. Translated into terms of visual experience, this would be like a semi-transparent door with extra light coming in around all the cracks. Not only that, but the way the light came around the edges and through the door would enable you to perceive clearly and in three dimensions what was happening on the other side of the door!

'Hearing' is not really the right word for this faculty, because we immediately relate it to our own highly specific sense of hearing. It is sonar 'visibility', the term to use is difficult to find, because of our own personal human associations, and because our human language relates to *our* mental and sensory experiences, not to that of a bat or a whale.

It is as if we were to carry our own sonic searchlight with us. That is, we could emit our own sound, somewhat comparable to emitting one's own powerful beam of light – something no creature seems to have accomplished so far, though the visibility of pheromones and other substances at all times of the day and night is at least partially akin to it.

To understand what other creatures perceive we have to do the well-nigh impossible. We have to get outside our human perceptions and enter into another world. We need to see things through *their* mind, through *their* sensory and motor structures – subtle and gross – not our own.

So these particular species of bat or whale are living and working in a sonar world. Just watch a bat weave its way through the evening sky and you may begin to glimpse something of the world it perceives.

In water, sonar is essential for long distance perception, for light cannot penetrate more than sixty feet or so. Those species of whale which feed deep beneath the waves, frequently possess weak sight, just as we humans possess pretty weak hearing and smelling, comparatively speaking. But their 3-D sonar perception is excellent. Some of the larger whales, measuring over 100 feet in length, may be quite unable to perceive their own tail flukes visually. But their acute hearing and sonar no doubt tells them exactly what is going on all around.

A sperm whale, for example, cannot see directly fore and aft, a factor made use of by the mariners of old, who quietly crept up upon them from straight astern or dead ahead. Their vision scans two quite independent fields of 120° on either side (see Figure 4.1). Can we imagine what it is like to have two non-overlapping visual fields? We cannot, because our human mind structure is not formed in that way, neither are our eyes.

Many whales also spend some considerable time swimming on the surface, where good sight is required but where their sonar is of little use. Clearly they are able to switch mentally between auditory and visual worlds. No small feat and something quite alien to ourselves, difficult to imagine. Under water it is mostly a sonic world and on the surface it is mostly visual.

For those readers who are optically minded, you might also like to ponder over the design difficulties associated with a lens system which comes into direct contact with water, where the refractive indices of the organic water-based lens and the surrounding watery medium itself are very similar. This is why man sees only blurred images under water unless he wears air-filled goggles. Yet whales and dolphins can see well, both in water and in air. An extraordinary feat of biological engineering.

The auditory organs of cetaceans contain immense air-filled sinuses as resonance and receptor chambers, as well as fluid-filled compartments, too. Both media have differing advantages and disadvantages for generating and receiving sound waves, and cetaceans are designed to reap the best of both worlds. Sound, for example, travels four times faster in water than in air, a useful bonus for the users of underwater sonar.

Actually, scientists thought for many years that whales had no sense of hearing because they had no external ear lobes like ourselves, a clear example of anthropomorphic projection. But whalers could have enlightened them in this respect, for they knew that whales possessed acute hearing, the constant 'twittering' of the White whale even earning it the name of the sea canary. But no doubt such observations were considered to be 'old wives' tales'. But 'old wives', as well as old sailors, can be great observers of nature and be well worth listening to. Ear lobes, of course, on a whale would pose problems underwater where streamlining is so important and in any event, they would not perform as efficiently in water as they do in air.

Sound production and hearing in the Bottle-nosed dolphin had been studied more than those of any other cetacean. Ken Norris and others (see Bibliography) believe that their sounds are produced in the nasal structures below the blow-hole. All cetaceans draw in fresh air through this nasal orifice down into their lungs, releasing it and drawing in a fresh supply whenever they return to the surface. So they always have a ready supply of air in their lungs with which to generate clicks and sing songs. But they possess no vocal chords as man and other mammals do.

So Norris suggests that in a Bottle-nosed dolphin the sonar clicks are produced just below the level of this blow-hole. The sound is then actually directed and focused by a reflector and a lens, as in an optical system (see Figure 4.2). Firstly, it is reflected and given direction by the cut-away skull bone, acting as a parabolic reflector. This beam is then focused by an acoustic lens, the *melon*, a fatty body situated right in front of the blow-hole and conspicuous by its presence in many cetaceans. Norris also points out that the complex system of internal air passages would prevent the production and reflection of random sounds – it is important for the dolphin not to confuse itself!

It is certainly verified experimentally that the sonic clicks produced are both highly structured and directional. So *Tursiops* is clearly

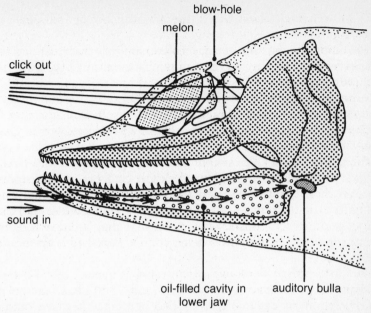

Figure 4.2   The accoustic sonar system of the Bottle-nosed dolphin
(Tursiops truncatus). Acoustic clicks are generated in the blow-hole,
reflected into a beam off the concave skull bone and focused by the fatty
melon, which acts as an acoustic lens. The returning acoustic signals are
channelled along the oil-filled cavity in the lower jaw onto the auditory
bulla. (After Norris and Bonner).

operating a form of acoustic laser radar. Something which man has
yet to design for himself!

But this is only half the story, for Norris also believes that the lower
jaw acts as an acoustic channel or probe for the returning sound. The
laterally located, inner ears of the dolphin are highly sensitive and
quite capable of detecting the sources of 'ordinary', environmental
sounds, thus building up a three-dimensional acoustic 'world view'.
But sonar echoes received from a narrow beam of sound directed dead
ahead would not be so readily received by laterally placed ears. The
sonar system is therefore a superbly designed 'addition' to an already
acute 'conventional' hearing system.

Norris points out that the lower jaw of the dolphin and many
cetaceans is actually comprised of an extremely thin and expanded
sheet of bone, possessing a minimum thickness varying from only
0.1 to 3 millimetres among the differently sized species. Between its
upper and lower layers there lies a cavity, filled with an oily material
similar to that found in the melon (see Figure 4.2). The returning sonar

thus passes virtually unimpeded through the 'auditory window', the delicate layer of tissue covering a thin lower mandible, travelling thence along the jaw until it reaches the auditory bulla, an auditory sense receptor lying close to the point of articulation of the lower jaw with the remainder of the skull.

Norris' ideas seem highly plausible and if this amazing and beautiful process is a correct understanding of part of the physical aspects of the dolphin's 3-D sonar system, it also explains why they move their heads from side to side while they are emitting their characteristic echo-locating clicks – they are simply scanning their targets for angular, 3-D information. It also explains why those cetaceans which produce no sonar clicks also lack a melon and lower jaw modifications. Many such whales, however, do produce powerful vocalizations and their lack of such structures does not mean that they are not equipped with some other, equally well-designed, auditory and echo-locating system.

The Bottle-nosed dolphin's sonar sense is remarkably accurate. Emitting clicks containing between 20 and 230 pulses per second, and lasting from around one to five seconds, they are capable of distinguishing different sizes of fish to an accuracy of a few millimetres. But not all species of dolphin possess such accurate sonar and, indeed, when the water is clear, even the Bottle-nosed dolphin will readily search for fish using only its eyes to guide it.

This does not mean that among those species not possessing *Tursiops*' style of sonar equipment, sonar is not of such importance to them. It is just that, as always, unless we can get a feel for the mental meaning and perception experienced by a creature in relation to its own world of sensory perception and motor action response, we can never really know how its own particular reality appears to it.

Considering the other sensory systems of cetaceans, they appear to have no sense of smell, for they lack any obvious olfactory organ, olfactory nerve or olfactory cerebral centre. Sonar, hearing, vision, touch and taste seem to be their major senses. Fishes, however, do have a highly developed sense of smell, as do many marine invertebrates. Smells travel well under water, though what the experience of underwater scents may be like is open to imagination. Since, under water, the 'scents' are in solution, smell and taste must be very similar. Furthermore, whales are supposed to be mammals who returned to a life in the sea – so, if the evolutionary theory contained all the answers, why should they have ever lost what to most mammals is an exquisite and finely tuned sense organ?

Having entered this discussion concerning the sonar world from a consideration of the different visual fields of animals, let us again pick up those threads. A rabbit has 360° vision (see Figure 4.1), a useful faculty if every predator around considers you good for dinner. But their range of binocular vision is only about 30°. However, *this is both in front and behind*. In front, this vision will help a bunny exactly locate its next mouthful, as well as accurately assess the character of the terrain and the location of its bolt hole when being pursued or when simply going home for a snooze. Rabbits often disappear down their holes at great speed and without slowing up, so they clearly need an excellent assessment of exactly where to jump, to avoid bumping into the sides of their front entrance.

But it is the rear-view binocular field which is of particular use in keeping the rabbit safe. If you have ever watched a rabbit being pursued by a dog you will notice that whenever the dog gets close, the bunny begins to weave this way and that to confound Fido's pounce. How does he do it? How does he know where the dog is? It is through being able to see simultaneously both backwards as well as forwards, without the need to glance over his shoulder. So, looking forward, he can see where he is going and, simultaneously looking backwards, he can see exactly where his pursuer is and what is the most appropriate avoiding action. In fact, I have seen pursued rabbits tilt their heads back, no doubt increasing their rear field of view.

But what is it like to possess 360° vision with two, narrow and separate binocular fields? To appreciate that, you would need to have a rabbit's mind structure, as well as his eyes.

Ducks, too, have 360° vision, with eyes on the sides of their heads, but with limited binocular vision. Owls, on the other hand, who also have exquisite 'binocular' hearing, enabling them to pin-point mice on a dark night, have excellent binocular vision as well, with eyes at the front. And they make up for a curtailed visual field by being able to rotate their heads through 180°. In fact, their necks are so mobile that they can keep their heads in the same position while a branch sways in the breeze underneath them. The body moves, while the head stays put! A most remarkable feat and quite essential if you are scanning the area for a likely meal. Try, for example, examining the detail of something several yards away while jumping up and down upon the spot. You cannot see things so clearly. Similarly, many hawks can hover with their heads almost motionless on the most windy and blustery

Figure 4.3   The cutaway snout requirements of laterally placed eyes, allowing the squirrel to see all around it. Bulbous jaws would hamper vision, and the ears, too, have to be high up and out of the way.

of days. Their bodies move in the wind while the head stays put.

Interestingly, the design requirements of eyes on the side of your head, which is where they need to be for all-round vision, plus the need for excellent binocular vision, leads to the cutaway snout or long face, such as we find in many rodents, including squirrels (see Figure 4.3). If all you could see were your nose, or the side of your face, nature's design system would clearly have slipped up somewhere. And this we never see. Binocular vision, or some other sense with accurate three-dimensional assessment, is essential for catching prey or performing acrobatic feats among the trees.

INFRASOUND, ULTRASOUND AND HUMAN SOUND[1]

Not all biologists accept that insects do communicate electromagnetically. Some prefer to think in terms of sound signals and the 'smell'

[1]Many of the factual details in this section were shown in the BBC TV series *Supersense*, and can be found in John Downer's fascinating book from the series, of the same name.

or 'taste' of pheromones. For myself, the evidence detailed in Philip Callahan's *Tuning in to Nature* does seem pretty conclusive. Certainly, few entomologists doubt that the amazingly intricate structure of moths' antennae are specific pheromone detectors.

But there is no reason why both these and other sensory systems should not be employed. There are, after all, many species of invertebrate, of which but very few have been studied in any detail. It is thought, for example, that the long and feathery antennae of male mosquitoes pick up the specific sound wave patterns generated by a female mosquito who is ready and willing for a sexual encounter. She vibrates her wings at a particular frequency as she flies, emitting a sound, to which the antennae of the male are exactly tuned, acoustically. They resonate like the string of a guitar.

Many sounds produced by living creatures, however, lie in frequencies outside our human threshold of hearing and are thus designated as *ultrasound* (higher frequencies) or *infrasound* (lower frequencies). Many creatures use these sounds for communication. Many of the sounds we hear emanating from insects, for example, are only the bottom end of a spectrum of sound signals much of which lie above the auditory frequencies to which our ears are attuned.

Indeed, such communications are not without the deceivers and manipulators, as in all walks of life. For some species of spider will blatantly mimic the 'come-hither' sound signals and even sex pheromones of female moths, luring lusting males to a dinner party they had not expected.

House spiders have often been reported to appear upon the scene when music is being played. I have a few which appear on my living room carpet whenever I watch certain TV programmes. This however is not due to a desire to rock'n'roll but represents an instinctive response to signals interpreted by their mind configuration as a call to food, sex or perhaps just good arachnid company.

Similarly, when a peacock butterfly opens and then suddenly closes its wings, it generates a sonic clap outside human hearing which helps to scare away predators. Rats and mice even use ultrasonic echoes to determine distances and sizes of objects in the dark – this is one of the reasons why they squeak. It tells them where they are relative to walls, obstructions and each other. What we hear as squeaking is only the lower end of a sonic signal mostly beyond the range of our human ears.

Many owls have such a perfect sense of hearing that they can locate and catch small creatures in the pitch dark, landing upon them with the claws correctly aligned to grasp them along the spine. Cats, too, are sensitive to ultrasonic frequencies and can hear mice and voles in hidden places. In fact, such sounds may be picked up in their whiskers, resonating like guitar strings, rather than by their ears.

Some creatures are sensitive to infrasound, to frequencies below our normal frequency range. The kangaroo rat is so aware of such sound that even on the darkest of nights it can hear the approaching slither of a hunting rattlesnake. And so finely tuned is its sense of hearing and its appreciation of timing that it will go on feeding until the very last minute, only springing to safety *after* the rattler has launched its otherwise deadly final strike.

Birds feeding on the road seem to wait until the very last minute before zooming off to safety, though inexperienced fledglings normally suffer a high casualty rate on our early summer roads. It seems probable that the high metabolic rate of such diminutive creatures is a reflection of what appears to us as a high-speed mental appreciation of events. Or maybe it is simply that a second to them is a vastly longer subjective experience of time than it is to us and other creatures, possessed of a slower metabolic rate.

Elephants communicate with each other almost continuously by emitting infrasound frequencies below our human auditory thresholds. Infrasound is far more penetrating than ultrasound. Their large thick ears are clearly required for more than mere show.

Conversely, the wafer-thin ears of many bats act as sound frequency filters allowing all but the most ultrasonic frequencies to pass right through. Some have hearing which is so acute that they can detect insects as tiny as a midge up to 60 feet away. Some birds also use low-frequency sound signals for territorial, sexual and other social communications. In fact, it is possible that the distinctive low-frequency sounds and hums generated by deserts, mountain tops and other locations act as homing beacons. For such sounds can carry for hundreds of miles.

Under water, the world of sound signalling takes on an additional significance since sound in water travels much further than light, moving a great deal faster than it does in air. The pistol shrimp possesses one of nature's most innovative designs. One claw is modified to snap shut with a sonic crack capable of stunning its intended prey. The effect on nearby fish is devastating. With its other claw, the shrimp then holds the disoriented fish and fires a barage of 'shots' at its head before settling down to dinner. The pistol shot is so loud it can be heard well over half a mile away.

Fish, too, often incorrectly thought to be incommunicative and largely undemonstrative creatures, even 'sing', like birds, and communicate territorial, sexual and social signals in the realm of low-frequency sound. And just as blackbirds perform a public service by letting off a loud 'chaka-chaka-chaka-chaka-chaka' whenever they spot a pussy on the prowl (resulting in the mass flight of all their avian friends into the taller trees), so too do squirrel fish, and most probably some other fish, sound off a penetrating and quite identifiable infrasonic burst upon the approach of a predator. Clearly there is more to such cross-species communication than a merely self-centred 'survival of the fittest'.

Fishes' 'songs', one should note, do not seem to be nearly so complex as those of birds, perhaps reflecting their lower position on the scale of consciousness and intelligence. So while the male bi-coloured damselfish attracts his mates by chirping – the sound increasing to a crescendo if a female happens to pass by – there is a species of minnow which purrs, the cod grunts and the love song of the male haddock resembles the sound of a motorbike, a quite unmelodious call which it can sustain for periods of up to twenty minutes.

And, as always, there are the terrorists who prey upon the daily needs and activities of other marine creatures. For sharks and other large predatorial fishes can hear these infrasonic signals in addition to the sounds created simply by the swimming activities of a shoal of fish.

AS MANY PHYSICAL REALITIES AS THERE ARE CREATURES?

And so it goes on. Every creature would have a different tale to tell if only it knew how. Every one has a 'strange' perception of reality, all its own.

Many four-legged animals have an intensely strong sense of smell. Perhaps smell can also be three-dimensional. Even we humans can tell from which direction a smell is coming. Or maybe the spectrum of smells is perceived in some way similar to our human perception of colour. Most mammals apart from man are thought to be unaware of colour, or the differentials between wavelengths in the light to which their eyes are sensitive.

Colour, like all sensory perception, is, as I have pointed out, very much a subjective, mental experience. The necessary mind structure is required as well as the requisite optical structure. So do our domestic dogs perceive smells with the same range of subtle 'hues' as we perceive colour? We certainly do not smell things in the way they do. Again, one needs their mind structure as well as the nasal physiology and brain organization, to know just how such an experience would feel. 'Red' is a subjective mind experience, not an outward experience, and so it is with all sensory experience.

Dogs and many other animals, especially hunting creatures, actually lead with their noses. Even on dry days they are kept moist, ready to dissolve and transmit the perception of interesting molecules.

If you have ever lived with a dog, you will know how important is his (or her) sense of smell. They smell each other, the cat, any other passing creature, all their human associates and anything else considered worthy of investigation. They leave scent messages under trees which then so fascinate other dogs who pass that way. They live in a world of scents and aromas which we cannot even begin to comprehend.

One whiff of a piece of clothing and the appropriate commands, and off goes Fido, nose to the ground, casting this way and that to pick up the same scent.

Like many other mammals, bats are also particularly well endowed with olfactory receivers and transmitters. They possess an impressive array of scent glands under nervous and endocrine control, the hormones involved being mostly steroids. Among the various species, such specialized glands have been found in almost all anatomical locations including the anus, sex organs, wrists, ankles, chest, under their leg articulation, upon their shoulders and back and in a variety of places around the head, face and ears. Furthermore, faeces and urine are used as markers of individual 'personalities'. At least one species of African bushbaby, for example, cups its hands and urinates upon them before setting off into the African night. The personal scent trail which it thereby leaves provides an easily identifiable highway home. Much better than a long piece of string!

A marked difference between the sexes in the structure or presence of such glands indicates that some aspects of sexuality are communicated by aromatic means but, as always, one would need access to an individual bat's mind to know how such messages are interpreted and what behaviour results. The world of multiple scents, like that of three-dimensional sonar is not one into which we humans can readily project our minds. It is only an extension of our human form of sensory perception.

Man uses his eyes for both long- and short-range sensory perception, but some creatures rely more on one sense for long-range and on another for short range perception. Dogs, for example, and many mammals, rely heavily on their noses for close-range perception, using their eyes to scan the wider horizons. If you have ever watched a dog running across a field, unless he is tracking something, he will often run with his nose in the air, using his eyes for long-distance viewing, but as soon as he reaches an area of interest, down goes the nose, far closer to the ground. Of those bats which use echo-location many also use sight to locate a prey at a distance, but employ their short-range sonar for actually closing in and catching it. Interestingly, many species of bat seem not to favour flight on moonlit nights, possibly because of the danger from owls and bat hawks. Their vision, of course, is well designed for use in what we humans call the 'dark'.

So all around us are individual physical realities, each overlapping the others, each different. And the difference lies primarily in the mind. This is true even among humans. We all see things differently, according to the karmas which determine our own personal mind set. But these difficulties of understanding are only reflections of the intellect's limitations. From a higher point of perception, it is all a natural part of the outworking of the greater Mind.

INTERLUDE

OTHER NATIONS

We need another and a wiser and perhaps a more mystical concept of animals. Remote from universal nature, and living by complicated artifice, man in civilization surveys the creature through the glass of his knowledge and sees thereby a feather magnified and the whole image in distortion. We patronize them for their incompleteness, for their tragic fate of having taking form so far below ourselves. And therein do we err. For the animal shall not be measured by man. . . . They are not brethren, they are not underlings; they are other nations, caught with ourselves in the net of life and time, fellow prisoners of the splendour and travail of the earth.

(Henry Beston, *The Outermost House*)

5. BEHAVIOUR PATTERNS

In the main, we have been exploring sensory perception and physical reality. But that is only half the picture, for sensory perception is always linked to motor response, to outward behaviour. In the fields of mental energy, these exist as the sensory and motor indriyas, as we have discussed. And the two are closely interwoven. There is no point in being able to locate something, either visually or through smell or sonar (whatever your métier), if you do not possess the motor organs and behavioural patterns to deal effectively with that something. This might be to chase it, grab it, attack it, eat it, live in it, avoid it, leave scent markings on it, make love to it or whatever.

So the sensory and motor indriyas of all creatures are closely integrated within the mind structure of a creature. Behavioural responses are thus instinctive, automatic, totally appropriate and well-designed, since holistic integration is the nature of all mind-body energy fields.

The sensory and motor indriyas are an integral part of the internal mind configuration of all living creatures. Their sensory perception as well as their motor responses – their behaviour – are thus totally consonant with their bodily form and function. They do what they have to do as whole entities, further integrated into the greater wholeness of soul, Mind and nature.

The ability to *learn* about the environment and even communicate that information to one's fellow creatures, as bees, birds and animals do, is dependent upon the inner completeness of the tattvic tapestry. But even the one-tattwa'd plants discover where light, food and water are to be found. They readily respond to these stimuli, though it would be difficult to call it learning. Their leaves, stems and branches orientate themselves to the light, while their roots seek out minerals and water.

That all living creatures possess a capacity to learn is apparent from the way they find their way around an environment. Even butterflies take note of being captured at a certain location and when subsequently released, avoid the area for some while after. Creatures from bees, butterflies, scorpions and spiders, through fishes, birds, mammals and man exhibit territorial instincts, so they must be able to learn the extent of their territory. Human trainers of all species appreciate the differing capacity of creatures to respond to training and sometimes to perform antics which they would never perform in the wild, though these are never completely outside their basic behavioural repertoire. Learned perceptions and actions have to be within the overall confines of a creature's mind structure. Lions may learn to jump through hoops or sea lions to clap their flippers while balancing a ball on their noses, but lions jump over obstacles in their home territory or laze around under a favourite tree, while sea lions are naturally dexterous, chasing fish with great skill. But you never saw the king of beasts clap his forepaws nor balance a ball on his nose!

Even new language, new communication signals can be taught. These may be auditory or of bodily gesture, but neither the response nor the signal will lie completely outside the creature's normal behavioural repertoire. The trainer's body postures, hand movements and words, whistles or calls are perfectly understood by creatures which are not naturally born with a mind set which includes such words as 'sit' and 'walkies', or whistles and haloos! But among themselves they have signs and sounds which indicate the same activities.

The birds and squirrels in my garden come for food as soon as they hear my whistle, even though the food may not yet have been put out for them. They come at the whistle, not the sight of food. Pavlov demonstrated this many years ago.

Body language is important, too. A calm stance breeds a sense of trust and security, while a threatening pose will make creatures run, just as a dog will cower or a cat will flee merely at the kind of look expressed upon your face, even in the absence of words. A sharp, aggressive movement of head, hand or shoulders is understood by almost all creatures.

Creatures, therefore, are not disconnected from each other, though only fellow members of a species appear to possess the capacity for the fullest understanding of each other. But appreciation of subtle energy fields and emanations, as well as visual, auditory, olfactory, gustatory, tactile and other sensations – all these can be a part of a creature's instinctive communication repertoire.

All living beings are linked at a mind level and all behaviour is an utterance, an 'outer-ance', an expression of the creature's inner being, just as it is with ourselves. Such behaviour may be specifically intended as communication or it may be only the natural expression of the individual's mental content of the moment. But the mind always expresses itself through the body, consciously or unconsciously, just as it does in humans.

A human mind dulled by its own self-absorption may fail to perceive this, but in stillness, this awareness automatically comes into our consciousness, without any effort of intellect or will. When we begin to understand the nature of the workings of our own minds, then we come to understand the manner of the mind's activity in all other creatures and, indeed, throughout creation, as the primary pattern-maker, the source of all the blueprints, the weaver of the web of energy relationships, the essential formative principle running across the pure face of consciousness.

ADOLESCENCE

Adolescence seems to be as much a matter of mental learning and maturation, as of physical growth. In man this is very clear, for although a child may be sexually mature and fully grown quite early on in his or her teens, they are still a long way from any degree of emotional maturity. The time of youth in all creatures is a time for exercising and learning to use the mental as well as physical apparatus with which they have been born, of trying out their preprogrammed patterns. The basic mind structure will always be there, but even instinctive patterns require training. The mental, as well as physical, muscles must be exercised.

A bird is born with an ability to fly, yet the mother bird will *urge* her offspring to take that first flight, calling to them from a nearby branch. Or a cat will bring home a live mouse to teach her kittens how to catch it. The kittens know instinctively what to do, they just have to learn exactly how to do it skilfully. Similarly, a mother cheetah on the African plains will let its young 'play' with a gazelle. Or a human parent will take delight in teaching their child to walk and talk and do so many things. Play is always geared towards learning, towards exercising the subtle and physical structures. This is why we should choose toys for our children with some considerable forethought.

One should note, however, that although a mother bird may encourage her chicks to fly by calling to them from a branch some distance away, holding some tasty morsel in her beak as an inducement, all the same, she is not actually *teaching* them to fly. She is only encouraging them to use an instinctive faculty they already possess, to develop the already existent, mental and physical muscles.

Perhaps it is this which is so charming in the actions and behaviour of the infants of all species: the freshness and the tentativeness with which the life within explores its new mental and physical apparatus. The unthinking brashness of the adult expert is altogether lacking in the youngster and our nurturing and helpful instincts are unconsciously aroused, wanting to help the recent arrival become familiar with its new physical overcoat, gently coaxing it into utilizing its full capabilities, like the mother bird calling from the twig.

A creature, therefore, can be taught nothing for which the inner, instinctive potential is not already present as a latent pattern in mind and muscle.

It seems to be a facet of nature that the lower a creature is on the scale of consciousness, the less there is for it to learn, and the more slavishly bound it is to its instincts. Insects, for example, do not teach their offspring, though wherever there is the ability to move about, there is still a capacity for learning details of the immediate environment. Bees, for instance, have an elaborate communication ritual – the famous 'waggle dance' – indicating the exact direction in which nectar is to be found. But this instinctive patterning is not taught. No bee *teaches* another bee how to understand and communicate their language. They are born with the ability: instinctively and automatically it is a part of their subtle mind structure. Adolescence in bees may include some learning, though no one but experience appears to do the tutoring.

Generally speaking, the higher the creature is on the scale of consciousness, then the longer is its period of adolescence, and the more is it reliant upon its parents, because the greater is the scope for learning. There may be some slight variations due to environmental circumstances or the physical size of the creature, but as a general principle, this is true.

Play is also an expression of what is described rather dryly by behaviourists as *social bonding*. It means life-to-life, mind-to-mind contact, enhancement of well-being, affirmation of being alive and contending with the vicissitudes of existing in this world. Among humans, social bonding includes everything from nodding, winking and shaking hands, to having a good cuddle. It also encompasses fashions of clothes and social custom. Other creatures, according to their own instinctive mind structure, have similar social contacts. It is all a part of life relating to life, mind relating to mind.

Considering the deeper aspects of being alive, one notices, too, that the higher the consciousness, then the greater is the capacity for both love and the awareness of other creatures as living entities. This arises naturally, because ultimately consciousness in its highest essence is also Love. So Love and Life are the same thing. Therefore, the less this inner consciousness or Life Force is hemmed in by an instinctive mind structure, then the greater is the capacity for caring feeling and the more an offspring is capable of learning. The capacity for learning, the need for tuition and the mental ability of the parents to care are thus all automatically manifested in synchrony. A beautiful integration within nature's processes, as always.

ANIMAL LANGUAGE AND MIND CONFIGURATIONS

Just as we find animal perception hard to imagine because we have a different mind structure to other creatures, so too do we have difficulty understanding their language. We do not have the same pre-programmed instincts and mind patterns, the same tattvic constitution, and we thus find it difficult to know exactly what is being communicated.

Some communications are obvious by their very nature or by the response which they elicit – the alarm calls of some creatures which other species automatically recognize, for instance. Or the barking of an excited dog, the purring of a cat, the bellow of an angry bull, the insistent buzzing of a hive of bees when disturbed. These communications many creatures can understand in broad outline. The alarm call of the blackbird or the chattering hiss of the squirrel when a cat walks into my garden tell all and sundry that danger stalks near at hand. And all creatures who hear it in time take their own appropriate, evasive action. But we do not readily understand the subtle nuances of meaning and expression which members of the same species can communicate to each other.

Man understands his kind of language. Whatever the language is, you will find names or nouns for objects, verbs for events, adjectives for qualifications and adverbs for linking these together coherently. All human languages have these. Even if we cannot understand the words of a foreign language, at least our minds are all human such that communication of meaning – *human meaning* – can take place.

Meaning is thus *prior* to words and language, it lies within the mind. Even without *verbal* language, we can convey meaning. But we do not understand the nuances of bird, animal, insect or other creatures' communication, because we do not possess *their* inward experience of mental meaning. That is a natural part of *being* that particular creature, of being surrounded by that particular configuration of mental and tattvic apparatus.

Language, whether verbal, bodily or of any other kind, essentially implies the conveyance of this *meaning*. When we speak to dogs, horses, dolphins, chimpanzees or to any other creature, we unconsciously employ gesture, body posture, facial expression, tone of voice, and so on. All these are a part of the conveyance of meaning.

Among humans, even when we do not understand the words, the manner of expression still tells us a great deal of what is being communicated, especially at a basic level of feeling, emotion and facts of physical life. It is easy to tell if someone is angry or friendly towards us, even if we do not understand their language. Words are only essential to put across more abstract concepts and intellectual ideas.

Similarly with communication between different species. We may not possess the same mind structure, the same subtle energy anatomy and function, but enough similarities are present for some sort of meaning to be conveyed. Fear, affection, surprise, greetings, farewells, pain, gentleness, aggression – all these and many more are readily communicated. Even 'Come here', 'Go there', 'Sit', 'Walk' and many other sounds made by humans can be linked by the training and learning process to actions and meanings quite familiar to the creature. But our dog does not understand the word 'sit' in the way we do any more than we understand the nuances of his barking in the way that he does. One could equally well train a dog to sit whenever we commanded, 'Fly'. It would make no difference to him.

This barrier becomes intensely noticeable when a member of one species needs urgently to communicate a matter of life or death to a member of another. Many are the tales of dogs, horses, elephants, whales, dolphins and other creatures who have spontaneously helped man in times of his distress or danger. Or who have gone to other humans to get help either for their own kind or for others with whom they are associated, human or otherwise.

This all implies a degree of creative intelligence and mind function which we largely overlook. But then many humans are largely unaware of their *own* mind, let alone that of others. Certainly it is uncommon for human perception to stretch to an appraisal of the individual character of a dog, a sparrow, a fish, an insect, a plant or an amoeba.

So, like other aspects of behaviour, while the basic capacity for its own particular language is built into the mind structure with which a creature is born, a certain degree of learning ability is inherently possessed by all living creatures. The more complete the tattvic configuration, then the greater the capacity for learning. But the essential mental parameters are already set, whether human or otherwise. The capacity to learn and respond is an essential aspect of life's natural expression. Life is drawn to life automatically. The God without seeks the God within, for He is within everything. Inwardly, all creatures are drops from the same Ocean.

So the jungle creatures may need to learn from experience that the alarm call of the langur and the chital hind denote the presence of a leopard, while the sparrows and small birds in my garden learn from experience that the alarm call of the blackbird indicates the proximity of a prowling pussy. Birds even have local accents overlain on the basic language patterns with which they are born. Such birds as starlings, which mimic many sounds from their environments, have been heard sounding off like factory sirens, ice-cream van signals or even police cars in a hurry. In the English Lake District I have heard them mimicking the curlew, a bird not found in the Cambridge area.

Descartes, who had a vigorously mechanical view about most things, believing the soul to be some largely discarnate entity entirely disassociated from the processes of living, saw animals through that same mind-filter. He insisted that animals were entirely mechanistic, that their behaviour was entirely stimulus-response based, with no space for any cognitive process. For Descartes, only humans had souls, animals and other creatures did not.

Darwin, on the other hand felt that if bodies could evolve from one form to the other, therefore the mind could as well, though he freely admitted that he had no ideas concerning the essential nature of mind itself, nor even of life.

These days, there is little doubt that creatures possess a mind structure. Many conventional behaviourists readily concur with the idea. The disagreement arises over whether the mind is the product of molecules and electricity or whether it is altogether of a more subtle nature.

A classic example of cognitive processes is that of animal memory and the demonstration of subsequent decisions based thereon. The experiment takes many forms, but one of the commonest is to show the animal subject a set of photographs – of trees, of a river, lakes, grassland and so on. The subject is then shown another set of photos in which some are identical to ones already shown, while others are not. In addition, the subject is provided with a 'yes' button to press if the photo is identical, and is rewarded with a morsel of food if the correct answer is given.

Creatures from pigeons to monkeys have shown themselves capable of performing this task successfully. Interestingly, the larger the set of photos, then the longer the 'thinking' time taken for the animal's response. Sometimes, as long as a minute passes before the 'yes' button is depressed by beak or paw.

But this kind of behaviour is readily observable in all wild and domestic creatures, who soon *learn* their way around their own territory, as well as *learning* the forms, faces, voices, scents and much more which identify their human and other companions. The squirrels and birds in my garden certainly know me individually. They will come to the window when they see me to request food – but not to those of my friends who are unknown to them. Any farmer or keeper of animals has observed this kind of behaviour.

In fact, a good friend of mine, Ruth Dayan, once studied a particular species of ant in the Israeli desert. This ant spits painful substances at the eyes of any potential trespasser on to its territory, but through slow acclimatization to her presence, the community she studied learnt to accept her presence without fear and therefore without attack. For many weeks she watched and noted their social behaviour without being spat at.

Whether it was a scent, or a visual or some other aspect of sensory perception, the ants came to know that the presence of this particular human was not harmful. Other humans who visited the site, however, were promptly attacked in the time-honoured fashion. And when Ruth transferred her quiet and gentle attention to another community some distance away, she was herself immediately attacked. A new process of acclimatization was again required.

This acclimatization, in one form or another, takes place when any living creatures meet. Almost all creatures have social and courtship rituals which affirm the nature of their personal association with each other, life to life and mind to mind. Humans shake hands, dogs smell each other, butterflies perform aerial gyrations together and so on.

What is different about human communication is the use of noises as symbols that represent a quite unassociated meaning. We call them *words*. The noises 'car', 'dog', 'house' bear no resemblance to the item in question, neither do the noises 'anger', 'fear', 'contentment' or 'meekness' bear any resemblance to the emotional states they are intended to convey.

Even more abstract ideas and concepts such as 'democracy' seem to be part of a thought and communication process unique only to humans. For while our dog soon learns to recognize his 'owner's' car or house, and certainly experiences fear and contentment, it seems unlikely that he could understand the meaning of democracy; though he certainly reacts to unjust treatment!

Human language is a faculty bestowed by the formative inner field of akash. Our language arises automatically from our ability to perceive our place in a world of time and space and – most importantly – to consider the implications of particular actions, that is, the faculties of foresight and discrimination. Verbs involve an appreciation of time; nouns of objects in space, or of mental concepts and experiences; adjectives of qualities between which we can discriminate; adverbs of associations in time, space and quality.

Examine the content and structure of any sentence in this or any other book and you will observe this same pattern. But without man's akashic linkage he would not have the instinctive perception of cosmic order enabling such mental processes to exist. The one comes automatically with the other. No other creature possesses this akashic integration at a mind level and thus no other creature has anything barely resembling our human kind of language structure. They simply lack the subtle energy fields or mental configuration in which such an appreciation of physical reality can arise.

Even the work of such pioneers as Sue Savage-Rumbaugh, teaching chimpanzees to recognize *lexigrams* (graphic symbols), simply underlines the fact that animals can only learn associations with particular sensory stimuli that mesh with their *already existing* mental structure. So learning that a particular symbol on a computer keyboard, situated in a particular room, delivers a banana from a feeder, is really the same kind of mental activity as *recognizing* the shape of a banana plant in a forest and *remembering where to find it again* the next day. But learning does, of course, imply memory and mental activity. All creatures have a mind, as we have said.

Thus, the pressing of a large computer-linked key identified by a pattern that the chimpanzee has been trained to associate with the availability of bananas, is no more surprising than its ability to associate the presence of bananas with the shape of a banana plant on which the bananas are hidden from view. Banana plants still indicate the presence of bananas even if the actual fruits are out of sight. And no doubt, the shape of a banana is linked in a chimp's mind with its pleasurable flavour, as it is with ourselves.

Our human tendency is to equate the *action* of the chimpanzee with the human kind of *mental process* which would lead to *our* pressing certain keys. Such anthropomorphism would appear to be unjustified, but it is certainly true that association is most definitely a *mental* process, whether we are dealing with insects, crocodiles, birds, beasts or humans. And the extent of these associative and other processes increases as we ascend the scale of consciousness. Man with his akashic element is the only species capable of perceiving the entire picture, such that, in mystic consciousness, he can even perceive the One in everything.

Like the chimpanzees, the ability of bluetits to learn the complex series of manoeuvres required to release a peanut from a special trick-feeder is no more difficult than many of the activities it learns during the course of its life, seeking out insects and seeds from the most intricate of places, getting itself and its beak into just the right position to extract the desired morsel. And no puzzle of human design can match the complexity and ingenuity required to build a nest – weaving the moss, twigs, pieces of hair and even lengths of wool and man-made fibres, collected in a myriad shapes and lengths, into a cosy den that no human could ever imitate.

Indeed, rather than attempting to teach a chimpanzee to communicate like a human, it might have been considered easier to teach it to communicate like a dog. But they can do neither, any more than humans can communicate like dogs or chimpanzees. The inner mind is simply different in its essential configuration.

No chimpanzee or dolphin has been taught the language of a dog or a human, other than those communications which relate to the basic mental features already common to their minds – territory, comfort, hunger, thirst, sex and so on. One could not teach a chimpanzee to *grow* bananas nor a dolphin to *farm* fish, for these are quite alien to its mind set and require the foresight, the ability to see and plan ahead, which only man possesses.

One might (perhaps) be able to to teach a dolphin to peel a banana, but only if it was incorporated into a sequence of actions required to achieve some goal considered *meaningful* to the mind of the dolphin – food, affection, play or whatever! So really, this is no more wonderful than a dolphin learning its way around the complexities of its natural marine environment so that it may find – and catch – the best shoals of fish.

Similarly, sea lions being naturally and instinctively dexterous may be taught to balance balls on their noses, respond to particular noises made by the audience or their trainer and even clap their flippers, but to them this is only a means of receiving food, affection, social contact and so on – all these being motives, goals and feelings they experience in the wild.

Again, none of the experiments in which chimpanzees have been raised, since infancy, entirely by humans has ever resulted in chimpanzees learning to speak like humans. Neither their mind structure nor their allied vocal chords can accommodate it.

Interestingly, chimpanzees have been taught to communicate using some basic signs from American Sign Language – a language of gesture and facial expression, both used naturally by chimps living in the wild. But even then, though it demonstrates the learning flexibility present in a chimpanzee's mind structure, no chimpanzee has ever communicated anything outside the natural instinctive content of its own mind. This becomes clear when one studies the essential aspects of chimpanzee behaviour and communication, both in the wild as well as after familiarity with human ways.

When chimpanzees were 'taught sentences' such as 'Mary give Sarah apple', using plastic symbols to represent each word, some animal behaviourists, as well as the popular media, always ready for a selling story, thought that a breakthrough had been achieved: that a chimpanzee had been taught to understand *grammar* – an intricate part of all human, conceptual thought processes, involving human perceptive abilities.

But more careful observation showed that the behaviour was simply an imitation of the trainer's actions, combined with an association of certain 'symbols' or shapes with particular things – nothing more than the association of the sight or shape of a banana plant with bananas, combined with a sequence of actions required to pluck the banana or to reach the location of the plant in the wild. Teaching a creature to perform a particular sequence of *actions* similar to human behaviour does not imply that its *mind* is then working *like* that of humans. The *mental activity* underlying the actions, still remains radically different. As the Indian saying goes, 'If a dog walks through a cotton field, he does not come out dressed in a suit.'

In summary, one can say that one cannot teach any creature to *mentally be* like any other, since we have no handle by which we may modify the essential mind structure of another creature. It is the same with humans. If we want to understand another person we have to learn how to see things from *their* point of view. If we insist that their mind patterns must match ours, then we are doomed to unhappiness and discord. Similarly, if we force them to outwardly behave like us, ignoring the fact that their inner mind still sees things quite differently, then we are living in a fool's paradise. But real understanding of the mind of another only comes when we have explored the inner, subtle energy structure of our own mind in deep meditation and mystic transport. It is an *experience*, not an *analysis*.

In fact, were we able to modify the mind structure of another creature, we would find that we had 'miraculously' modified the body as well. Though this is impossible to us as humans, nature does it all the time. But it requires the death of the old body and the recreation of a new one out of the mind-seeds of the old subtle structure. This is the law of reincarnation. We get a new body, mind structure and destiny which exactly matches our mental impressions, desires and tendencies. Nature is quite exact. And so the creation continues. It is nothing but a world of Mind.

Even among the members of a particular species, no two are alike in all respects. Many, if not all, birds and animal individuals have distinct voices, just as we do. If they are not a vocal species, or even if they are, their smell or other characteristics may also be different.

If you have ever stood on a sheep farmer's hillside, as I have done many times in the Western Highlands of Scotland, and listened to the sheep calling to (communicating with) each other, you will immediately notice how their intonations are all different, as indeed are their faces.

A mother penguin can find her very own youngster in a colony of several thousand, even after many days at sea, having left the young one to mill around on the ice with thousand of others which to us appear to be identical. Similarly, gulls and probably all other birds have the mental ability to distinguish their own young, their mates, members of their social group and so on.

A wren sings a song with twenty or more distinct notes per second. And these notes vary. Clearly, it has a mind which can actually *distinguish* twenty notes per second while singing them, also hearing the nuances and variations in both its own songs as well as fellow members of its kind. There is no point in singing a song indicating your personal territory if the songs of all your fellows sound the same!

If bees can communicate distances and directions, corrected for wind variations, by their waggle dance on the side of the honeycomb, then may not the birds and other creatures be communicating to each other, in some considerable detail, those things relevant to *their* personal realities? Things which we do not have the mind structure and therefore the relevant sense of mental meaning, to understand?

Mostly, as humans, we communicate details of our day-to-day experiences to each other. Gossip, plans, hopes, fears, problems, desires, and so on. Even politics, philosophy and science come into the same category. We talk about the world according to our human mind configuration.

So presumably, other creatures are simply talking and communicating about the world according to their particular mind configurations. It is limited in comparison to our human potential, but it is nonetheless communication. Other creatures, according to their own tattvic tapestry, will have communications to make to each other which reflect that inward condition.

So, if man has a mind and consciousness, then so do all living creatures. The inability to enter into an understanding with them is due to our limitations, our lack of perception as to what we are and what this 'world' really is.

THE SONGS OF THE HUMPBACK WHALE

The songs – for such they may truly be called – of the Humpback whale have intrigued and fascinated man for many years. Such songs, as might be expected, contain characteristics identifying the individual. There is also considerable repetition of themes. Each song consists of a constant number of these themes, each theme being made up of a number of phrases. No theme is ever omitted, but one of the individually distinctive aspects of the song is the number of times that a particular phrase is repeated. Such phrases may be repeated many times with slight variations in each repetition, so that the phrase at the end of the sequence is quite different to that at its beginning.

Variations on a phrase or a theme are common elements of our human musical compositions. But what are the Humpbacks saying to each other across perhaps hundreds of miles of ocean?

Humpbacks are thought to be mostly solitary creatures, though such songs indicate that close physical proximity is by no means necessary for social contact of a vocal nature. Certainly it must help them locate a mate and induce a feeling of social togetherness.

But almost all creatures communicate with their own kind, often – especially among the higher animals – in sheer play and frolic. And may that play also include their song, like the playful growls and snarls of puppies and kittens, or even of their parents?

And are some Humpbacks considered better songsters than others? Are some renowned amongst their fellows for the quality of their underwater ballads and harmonies? Are they even requested, by suitable cetacean encouragement, to 'Play it again, Sam', as an item of pleasurable entertainment for the Humpback population? After all, a dog or a cat will come for the rub and tickle it so obviously enjoys. So do Humpbacks have ways of expressing the same request for the repetition of a pleasurable sonic experience? We may never know.

INSTINCT AND LEARNING

Being equipped with some kind of understanding as to how other creatures are put together gives us an excellent handle for comprehending how and why they behave as they do.

Mind works by habit. Once established, even man finds it difficult to change his trend of mind, even if he is aware that it is leading to his downfall. Instinct is simply a pre-programmed mental habit, setting up the pattern for the behaviour of a creature. It comes along with taking birth into a particular bodily form. A chick knows how to break out of its egg. A young herring gull taps its mother's beak to trigger her into regurgitating its next meal. The young of many species take cover as soon as an alarm call is given. They instinctively understand the call and know, too, how and where to take cover: some below ground, some by crouching low and mimicking the earth or leaves, others flee to mother's wing or pouch, baby crocodiles take cover in the gentle jaws of mother's mouth. This behaviour is not learned, neither is its pattern to be found in the DNA.

Despite the breeding of dogs into such a variety of shapes and sizes that an alien visitor to our planet would be forgiven for thinking them all to be different species, yet these miscellaneous varieties all exhibit the same essential behavioural tendencies as their wolf ancestors.

Puppies will herd hens in a farmyard, just as a pack of wolves will encircle an ailing prey. Licking your companions is a form of social bonding, practised by many wolves and wild dogs, not only by our domesticated dogs. Barking, whining, snarling and whimpering are all forms of communication practised by Fido's wild counterparts. Tail-wagging, rolling on their back and exposing their tender neck, the tail-between-the-legs routine, the self-confident bound and posturing of the top dog – all these and many similar body gestures of greeting, submission and dominance are common to many related canine species, whether wolf, hyena, dog or jackal.

Characteristics may be refined by breeding to some relatively small degree, but such behaviour can never be fundamentally altered, any more than a dog can be bred into a cat, or a pigeon into a parrot.

There is a hidden subtle dimension to all creatures that makes them what they are, possessing the basic behavioural characteristics of their species. Man may learn language and perform many other normal human functions, but the basic abilities to be able to talk, to be a human, these come gratis. They are our human instincts, parts of the beings we are.

Nobody teaches a spider how to weave its web, a caterpillar how to build a cocoon, or a bird how to find the right building materials and construct its nest. Nobody teaches a child the actual mechanics of producing word sounds or how to use his mind to think or experience emotions. These are all parts of deeply habituated mind structures, pre-patterned and pre-programmed, parts of the formative field which underlie the bodies and the instinctive characteristics of all living creatures.

Even man, when living through habituated mental patterns, away from his true princely abode at the eye centre, behaves uncontrollably and predictably. The thief remains a thief unless he can change his innermost mental habits. Politicians rarely change sides unless their instinct is purely one of self-preservation, human personalities remain predictably the same. Criticism and cajolery are of little help in changing the basic trends of an individual's psyche.

A robin will attack a bundle of red feathers, a socialist politician will verbally attack a right-wing conservative, a spider will attempt to tie up a vibrating tuning fork if the tip is placed upon its web. All are indicative of habituated mind energies.

The less the intelligence of a creature, the more instinctive are the behavioural patterns programmed into it and the less it possesses the capacity to learn from experience. The greater the linkage of the subtle tattwas into the mind energies of the creature, then the greater its intelligence and the more complete its potential for awareness.

With the patterns within responsible for generating the patterns without, we can clearly perceive the process by which each creature is a whole. The subtle energies comprising the instinctive mental patterns automatically produce a body and sense organs to match. They become part and parcel of each other. The subtle, instinctive mind pattern of an insect may be looking for the particular infrared, pheromone emission which signifies the presence of his lady-love. His antennae are tuned to receive just that very emission.

It is a pattern-matching exercise. Mentally or instinctively he is looking for the pattern and his antennae are an outward expression of that inward instinct. Or he may be looking for a particular foodplant, or flower for nectar, searching for a special smell or a particular electromagnetic emission. Or he could be following a highway of pheromonal signals laid down by fellow members of his species – a trail leading to food, maybe.

Similarly, a migrating bird could be seeking a particular patterning of the stars or an instinctive orientation to the Earth's magnetic field. But nowhere in the genes does it say, 'Fly south, my friend, when the evenings draw in.' Yet there may be genetic information encoded on the construction of the particular physical sense and motor organs required to perform the deed.

The DNA, the genes, the cellular centre of biochemical 'intelligence' and focus, is only a part of the outworking of the process from within to without, of how the subtle energy blueprints become the outer forms with which we are familiar.

SOME MORE INTRIGUING FACTS ABOUT WHALES

Even relative to their body size, cetaceans possess a large cerebral cortex. This is the outer area of the brain usually associated with the elaboration and co-ordination of sensory and motor function at the physical level and reflecting the content and structure of the more subtle sensory and motor indriyas within the mind of the creature.

The mind functions associated with intelligence (memory, emotion, figuring things out and so on) – instinctive or otherwise – are more central to the life or existence of any creature while the sensory and motor functions are more peripheral. One does, for example, still have a sense of being, identity, thought, emotion, intellect and memory even in the absence of all sensory input or motor activity.

But in the absence of the central sense of being – that is, in unconsciousness – the sensory and motor functions are quite inactive, even if the sense organs are still conveying messages to the brain. And, interestingly, this arrangement is reflected in the cerebral cortex being peripheral to the more inward mid-brain, the physical focus of mind and consciousness. I discussed this phenomenon more fully in *The Web of Life*.

Whales and dolphins are largely carnivorous, but feed mostly on much smaller creatures – fish, lobsters, krill, plankton and so on, depending upon the species and where they happen to be in the world's oceans. The Killer whale is an exception, preying on seals, sea-lions and some of the smaller cetaceans – porpoises and dolphins, for example. Killer whales evidently evince feelings of considerable fear in these creatures for even the usually timid Grey seal has been known to jump right out of the sea, joining a man upon a rock, when Killer whales were about.

Like many carnivores, Killers seem to be adaptable, taking whatever prey is available and are not averse to salmon if they find them massing in an estuary prior to spawning. Even penguins and other sea birds can be included in the menu.

Since seals, dolphins and porpoises can cause severe damage to fishermen's nets in addition to scaring away the fish, the fishing industry have recently taken to playing underwater recordings of the Killer whale near salmon and pilchard nets to keep away such marauders.

This, however, may result in the seals coming to regard the call of the Killer whale as a dinner gong – with potentially fatal consequences. But in our present context, it raises the question as to whether the call of the Killer whaler is *recognized instinctively* by a new-born seal or porpoise or whether it is *learnt* during adolescence, while in the company of parents.

The individual sounds of other species may not be so readily recognized purely from instinct, though it seems highly likely that the unborn infant, while still in the mother's womb, especially in its latter days, would be able to hear such external sounds and be aware of its mother's response. Learning may thus begin even before birth.

The general framework of a creature's emotional repertoire is instinctive, not acquired, though the external triggers to emotion may be learnt. So it seems likely that just as a youngster does not need to *learn* to be afraid or to feel secure, they also respond instinctively from birth to the underlying *mood* of the call – like the *predatory* roar of a lion, the *angry* bellow of an enraged bull, the *contented* moo of a grazing cow or the eminently *social* bleetings of sheep, keeping the flock together. Since the call of the Killer whale is no doubt heavily impregnated with its hungry mental intention, these would be understood instinctively and automatically by the young seal or porpoise, since those feelings are also a part of their mind structure. This would be just like we ourselves can tell when a dog, or any other animal, is feeling in a dangerous mood without understanding the greater detail of their language. Such 'greater detail' would include things like *recognition* of the individual dog by its own voice, the *depth* of its excitement and so on, just as it would with ourselves.

Feeding behaviour among cetaceans is also of some fascination and there is clearly social co-operation between members of the same species, especially among Killer whales, which tend to hunt in packs, as do lions or wolves, using particular strategies for catching their prey. This, again, is a mixture of instinct and learning. Humpback whales have even been seen to weave a snare of air-bubbles – a bubble net. The Humpback, swimming a little way below the surface, emits a fine trace of bubbles from its blow-hole as it describes a circle or a figure of eight around a shoal of fish. A glistening screen of these fine bubbles, rising up through the water towards the surface, forms a vertical and enclosing 'net' around the fish, like an enormous tube, and the fish being disturbed by it, gather towards its centre. Moving underneath them, the whale then rises, gulping them down in one leviathan mouthful.

BUILDING A HOME

Have you ever really examined the amazingly intricate construction of a bird's nest? Each species builds in an individually identifiable fashion. Some build out of moss and leaves, some weave a hanging basket of grasses and twigs, others play the part of the potter, mixing mud and water to build beneath the eaves of our homes. Others line their nests with mud, with moss, with leaves, with feathers. Some dig burrows in sandy banks, others hollow out holes in trees. Some lay eggs among the stones, so camouflaged you would mistake them for the rocks themselves. Yet no one teaches these birds how to find their building materials, how to prepare them, how to weave them.

No creature ever deviates from its instinctive mental programming. The swallow always builds in mud, British song-thrushes always mud-line their nests, the weaver-bird never experiments with alternatives to hanging baskets.

Examining so intricate a nest one wonders how such a thing could be possible. How can the creature be so clever as to weave with beak and feet alone a tapestry no human fingers, however deft, could ever hope to imitate? Do we really believe that such an instinct is to be found encoded only into molecular and electromagnetic patterns within the brain? That such behaviour is somehow encoded into DNA?

A home, for any creature, is a place to live, to shelter, protected from at least some of the vicissitudes of existing in this world. Man is not alone in his architectural abilities. In fact his skills are far surpassed in many ways by his lesser cousins. Creatures from single-celled amoebae, spiders, insects, birds and mammals all build homes or shelters for themselves, often of extremely complex design.

There are amoebae which gather sandgrains of particular sizes – a process clearly requiring a *selection* procedure, an instinctive ability *to make decisions* – using them as a protective outer layer. Some secrete their own granules. Either way, when seen through an electron microscope, the result is often one of startling and beautiful variety in miniature. Some species of these single-celled creatures even select and collect tiny grains of sand which, rather than being used in the 'parent's' outer shell are collected together internally and passed on to the 'offspring' when the single-celled organism divides. There they are immediately employed by the 'youngster' in the formation of its new shell. Even amoebae make provision for their young.

Every creature builds by instinct. Some learning, some development of skills, may be present, especially among higher animals, but the basic and intrinsic format is always present, instinctively, as a part of the basic mind structure or instinct of the creature. Even man's dwellings – whether palaces, thatched mud cottages or high-rise blocks of concrete flats are only variations upon the *human* theme of a ground-based structure possessed of walls and a roof. We do not build like other creatures, but always like humans. We have human instincts, not those of other creatures.

The social communities and nests of bees and termites, the aquatic homes of sand grains and sticks built by caddis fly larvae, the webs and traps of spiders, the nests of some fishes, the huge pits over three feet deep and six feet in diameter of the green sea turtle, the ramifying burrows and nesting chambers of wood mice, the holes and domestic quarters of foxes, badgers, moles and rabbits, the amazing engineering feats of the dam and lodge-building beaver – all these and many more tell of an inner instinctive pattern of mental motivation lying beyond the realm of brain cells alone.

When scientists perform experiments which demonstrate that brain damage causes disruption to animal behaviour patterns, this does not indicate that the source of instinctive behaviour lies in the brain. It only shows that the brain is implicated in the process by which the inner mind structure in the subtle energy fields seeks outward expression.

If we unplug a TV set from the mains electricity, it ceases to function. But I cannot then say that the real source of electricity is the socket upon the wall. A detailed analysis of wall sockets will not reveal the true nature of the electron nor the source of the electrical power.

In fact, however far back I trace the source in a horizontal manner, I never reach an ultimate answer. Electricity does not arise from the wall socket, nor even from the power station. No power station has ever *created* an electron, the source of our domestic electricity. Even physics does not understand the nature of an electron and electrical charge. Though our modern technology and civilization is based largely upon our ability to harness the electron, conventional physics still has no real perception of what it is we have got our hands on.

Similarly, a perception that the brain is involved in the outworking of mental processes does not mean that the mind arises from the physical brain. There is a subtle heirarchy of inward energies. The fact that LSD induces spiders to build crazy webs only means that their brain is involved in the outworking of instinctive mind patterns – not that instinct is lodged in the brain itself.

There is no end to the recounting of social ritual, communications, behavioural patterns, sensory perception and motor activity, to be found among the creatures of this physical world, human or otherwise. Books have been written on just one aspect of one creature and yet have never reached the end of the story.

MYSTERIES OF MIGRATION

Far up in the northern reaches of Canada roam vast herds of caribou, a deer which we in Europe call the reindeer. Yet, 'No-one knows the ways of the wind and the Caribou,' runs the ancient Inuit (Canadian Eskimo) proverb. Signalled by changes in the weather, great herds of these deer follow ancestral migration routes to sheltered valleys and more ample food supplies. The routes have been trodden by generation after generation of their kind, for longer than man can remember, with only the slightest variations.

But what instinct, what subtle pattern of mental energy, drives the deer to seek refuge in this way from the advancing winter? How did the caribou first find their winter refuge? How is the knowledge transmitted from generation to generation?

One could suggest that there is a learning of the young from the old, that they follow scent markers from a previous year. Maybe they do, but how was the route first discovered? How did the first deer know just where to go?

Among birds, the mystery of migration is deepened even further, for the young swallows, born one summer in Europe, fly home to Africa *before* their parents. The pattern they follow is instinctive and also specifically geographical. How can such geographical behaviour be transmitted through molecules of DNA? DNA is only a *part* of the total patterning process of a creature, not its *only* formative mould. What, after all, patterns the DNA? For creation is a *continuous* process, a proliferation of being, from within-out.

Some of these caribou have in recent years suffered greatly, at human hands. In search of hydroelectric power, man has dammed one of the great rivers of northern Canada, flooding the area above the dam, creating swollen rivers where once were fords, used by the migrating herds. But the caribou knew nothing of man's activities and following the ancient trail they found their way barred. Clear confusion reigned in their instinctive minds, and while they dithered on the banks, the water rose behind them, cutting off retreat. Thousands of caribou perished in just the first year of this dam.

Or, consider the lemming, that famous little vole-like rodent and – in particular – the Scandinavian variety. When their numbers reach a peak, like bees in a hive, they decide to migrate to better pastures. Many do so successfully, along ancient trails, a knowledge of which is found only in the instinctive mind level of the lemmings themselves. But some, following one of these ancestral pathways, find that the land stops, unexpectedly. They find themselves upon a cliff-top.

Like the caribou, they cast about, this way and that, clearly in confusion. Their mind patterns do not contain a programme to deal with cliffs where once was a continuation of solid land. After a while, the instinctive urge too strong to withstand, they cast themselves off the cliff into the sea below. That is no great thing, for the lemmings can swim. Then off they go, into the wide sea, following the instinctive beacon which tells them to travel in that direction.

In ages past, the land must have been continuous, or at least the sea passage but a short one. Now there is nothing but the great ocean. So they perish. It is not a mass suicide as was previously and naively thought, but a migratory instinct which includes a specific geographical map.

Then there are the 'freshwater' eels, who spend most of their life in the rivers of western Europe. But they are not born there. Each year, eel larvae, about three inches in length, appear in the *salty* seas off the European coast. Here they undergo a most remarkable transformation, a metamorphosis into young, cylindrical elvers which seek out the *fresh* waters of the rivers and streams. Slithering through damp meadows at night, wriggling over rocks and up waterfalls, even finding their way into mountain streams, ten thousand feet up in the Alps – in these resting spots, they settle down and live for many years. But they do not breed. For they did not start their life here and their ancient breeding grounds are far away.

The larvae were hatched three thousand miles away in the Sargasso Sea, lying in the Atlantic between the West Indies and the Azores, containing a superabundance of seaweed of the genus, *Sargassum*. But their parents were not of European origin. They come, it is thought, from America.

There, in the spring, they lay their eggs at a great depth and the minute larvae float up to the more surface waters with the onset of summer. Here, they are taken by the ocean currents into the Gulf Stream, in which three years pass as they slowly drift towards Europe and make their homes in the rivers, lakes and pools, far distant from their place of birth.

But even here the story does not end, for after fifteen or twenty years, the urge to breed and migrate once again comes upon them and down to the river mouths they proceed, slithering over wet meadows by night until they reach the greater river, lying up by day in damp holes, enough water remaining in their gills to enable them to breath.

And when this urge to migrate and breed comes upon them, their bodies undergo some considerable changes. Putting on weight as fat deposits – food and fuel for the long journey ahead – their entire bodies become covered with mucus in preparation for the miles of slithering travel.

On reaching the ocean, they disappear. Marked eels from Europe never reach the Sargasso sea, only their counterparts who had taken a shorter journey to the American mainland. On entering the sea, their anuses close up, rendering them unable to feed, requiring that they live entirely off their own fatty deposits. But these resources are not sufficient for them to reach their migratory goal and it is assumed that they die before the journey is complete.

There is little doubt that breeding is their intent as they head out to sea, for their gonads have become fully developed. But what becomes of them is a mystery, for none has ever been found, even in deep Atlantic.

On the face of it, it would seem that the eels could breed in many alternative places, well within their reach. But the migratory instinct and the geographical pattern seems so deeply etched into their inward mental fabric that their pre-programmed habit rules the day.

Maybe, like the lemmings, the pattern dates back to an era when their goal was closer and the swim was not so long. The Sargasso Sea is also said to have been the site of the ancient lost island of Atlantis where the eels could have once again found fresh water and been able to feed after spawning in the nearby sea. For now it seems that even the eels from America spawn, and then die, in the Sargasso Sea.

But, like the lemmings, how does such a geographical knowledge become an inherited ancestral heritage? And then – as the aeons pass and the continents, oceans, mountain ranges and terrain all drift – why are not so many more migratory creatures found to be wildly adrift in their navigation?

The problem is easily stated: the instinctive mental fabric of a creature is related in specific detail to its environment. No creature learns its innate skills. The swallow building nests of mud, the archer fish shooting down flies by spitting water distances of up to *fifteen* feet – *every* creature, without exception, has such intriguing inborn capabilities, if we study them.

And just as modern science has no answer to the riddles of human brain and mind function, so too is there no satisfactory answer to the way in which a creature's brain, body and instinct are put together. Biological science cannot tell us of the mind of a creature any more than it can understand the nature and functioning of our human minds. Behaviour and DNA are uncomfortable partners, for mind functions cannot be so readily consigned to a molecular coding.

This is riddle enough. But many migratory creatures have such precise geographical instincts that one wonders just how so detailed a knowledge of local terrain can be transmitted from one generation to another without the involvement of any learning process.

Wafting vague answers about DNA are totally inadequate in the face of such mysteries. Are we saying that route maps can be encoded into the DNA of just one germ cell? How? DNA structure and mental abilities are two radically different things. Mind is not to be found in molecules any more than the works of Shakespeare were to be found in his genes.

The only conceivable answer is that there is more to life than molecules; that behind the sensory level of experience, we are dealing with a complex and dynamic tapestry of Mind energy. And that within that Mind – of which all individual minds are a part – there are 'laws', processes and relationships, the ramifications of which we have barely glimpsed.

Outwardly, we see only the effects, the image on the screen, of this vast sea of Mind. The patterns and rhythms of the sensory realm are only reflections of more fundamental attributes of Mind. Even the cycles of this world are also of the greater Mind. If therefore, there were cycles within the greater Mind itself, geared towards a regulation and administration of life upon our planet, then we can begin to see how enigmas such as apparent geographical inheritance can be understood. And how, as the continents and oceans shift and slide, so too are creaturely minds and instincts automatically adapted, for they are all linked to that same sea of shifting Mind.

So could there be, in the Mind itself, an autumn when old habits are shed and a springtime when old growth is renewed and new shoots flourish? Just as we have the ebb and flow of seasons, each with its own appointed purpose; just as the rest of night follows the activity of day, so too could there be vast seasons of the Mind, spanning aeons, which hold the world in balance, removing dead wood, and vitalizing and renewing harmony between all forms?

This is an interesting subject, tackled more fully in *Natural Creation . . . Or Natural Selection?*, but is introduced in the next short chapter where we take a brief diversion to present something of the vast integrated cycles by which life is maintained upon our planet.

6.   CYCLES AND SEASONS

NATURAL CYCLES

Time is a window through which we spy the world. And being engrossed in our present moments, we fail to grasp the significance of the time behind and the time ahead. We live in the moment and the greater cycles of nature span periods too vast for us to contemplate. Yet they are as important to life here as night and day, as winter and summer, as birth and death. But how do we get some feel for spans of time stretching into millions of years? It is difficult enough to envisage our own advent and demise.

If time were as a looped conveyor belt, endlessly revolving, with past, present and future all existing simultaneously, our constrained human view would be of just one thin sliver across this belt. How then could we ever understand the nature and functioning of the whole belt? The narrow view is that of our individual human mind. But the greater Mind is the creator of time and to reach the level of the Universal Mind within oneself is to view the whole conveyor belt, turning simultaneously.

That is a far off mystic goal, perhaps, but let us at least attempt to stretch our minds, looking briefly at just a few of these cycles, and of the indispensible parts they have to play in the maintenance of planetary life.

Mystics of past and present have written that there are seasons, at the physical level, in the manifested power of the Life Force, the Creative Word. Just as springtime possesses a vibrancy of fresh life unsurpassed by any other season, just so – say the seers – is there an ebb and flow of the Life Force, but spread out over vast aeons of time.

On a time-scale which may not coincide exactly with our present scientific calculations of geological time, they say that the cycle

commences with a period of 1.7 million years when the power of
the Life Force is immeasurably stronger and more evident in physical
affairs than it is today. The harmony is such that man is even said
to live, in a much etherealized condition, for 100,000 years. In the
Sanskrit *Puranas*, this age is called the *Sat yuga* where 'Sat' means
'truth' and 'yuga' has the same root as our word 'age'. It is thus
the age of truth, the age when life is clearly seen for what it is and
man knows his role in the totality of things. Opinion is replaced by
experience and direct perception.

This may, at first sight, seem like too much to swallow, though even
the Greeks and Romans of the classical era also entertained such ideas,
derived from their own mystics. But the power of this strong life-giving
vibration is said to wane slowly over a 4.32 million year period, until
man even forgets that he is really alive and conscious. It is through
this phase that we are currently passing.

Modern scientific man thinks that he is no more than a chance
arrangement of dust and water, of molecules – though he may
acknowledge that the atoms and molecules, indeed all life forms,
are highly ordered and organized! But the materialistic view has
no idea as to how this dynamic complexity in form and function
comes into being and is maintained. 'Self-organization' is the current
buzzword! Yet no-one can say whence this perpetual motion and
apparent 'self-organizing' capability has arisen.

So the clarity and vibrancy bestowed by the immanence of the Life
Force during a Sat yuga declines, giving way to ages of progressively
lower consciousness, until we reach the *Kal yuga*, where we find
ourselves today. 'Kal' means time, death, Universal Mind or the
Negative Power. Here, all life is at its lowest ebb and man has lost
all sense of direction. The Divine purpose of life is almost entirely
forgotten and he spends each lifetime self-centredly chasing after
vanishing mirages. The condition of the physical universe deteriorates.
Even the spiritual values of life are held in little esteem.

Ultimately, the balance becomes intolerable until, in what must
be akin to the most incredibly vibrant springtime one has ever
experienced, the world is flooded once again with the power of
the Life Force, of consciousness, of universal spirituality, devoid of
ritual, dogma and the outward trappings of religion. Whether this
change comes about within one year, over a thousand year span, or
perhaps over the weekend – I do not know, though geologically all
such time-spans are short periods. But it is said that the Kal yuga gives
way to the Sat yuga when the manifested power of the Life Force has
reached its lowest ebb.

One wonders, then, what must be the effect of such an influx of fresh vitality upon biological processes? Do we have here a key to understanding how lemmings, eels and other creatures gained their instinctive knowledge of the changing planetary geography? Perhaps at such times there is an opportunity for a general biological and genetic reorientation? The density or grossness of physical life being relaxed, the more subtle components become capable of realignment to the prevailing circumstances, all within the formative and integrating influence of the greater Mind.

One thing is certain – that a Darwinian-style, materialistic and linear, reductionist thought process cannot begin to describe, let alone explain, the integrated, complexly woven cycles and patterns of behaviour and physiological process that we see around us.

Nature is replete with cyclic processes. Our reductionist, analytical thinking attempts to form them, conceptually, into discrete linear pathways, but nature knows nothing of such superficial imaginings! She is one integrated whole. It is our minds which are divided. It is out of this divided and constrained perspective that we have developed the materialistic theory of evolution and try to fit everything into the confines of such a bleak and narrow view.

Our limited perception of time has difficulty in imagining events and cycles dating back just a hundred years, yet archaeologists have germinated and grown wheat from grains four thousand years old, discovered in the tombs of the Egyptian Pharoahs.

Bacteria which normally divide and reproduce themselves every twenty minutes can also form themselves into environmentally resistant spores capable of lying dormant, even entombed in rock, for maybe millions of years before a change of circumstance permits their release with continued growth and reproduction. How can a cycle covering such an immense time-span have ever evolved out of random mutational steps?

One whole phylum of creatures, the *rotifers*, seem to fit into no known evolutionary tree at all. They appear, like miniature miracles, in any pool of water, however transient it may be, complete with wheels or crowns of cilia, minute beating hairs which function equally well for both finding food and for locomotion. When the pools dry, the rotifers turn into minute particles of dust, wrinkled and desiccated, awaiting the next rainfall.

They honour no geographical boundaries and the same species are found in the pools of Africa, Asia, America and Europe. These spores have even been found high up in the atmosphere at 50,000 feet. 'For instant rotifers, just add water,' comments Lyall Watson, in *Supernature*, adding that there seems to be no reason why these desiccated specks should not fly higher, even finding their way into space. Perhaps indeed, he suggests, this is where they came from, born through outer space from some other planet, in some other star system. How does one study the life cycle of an organism that may live for millennia, or even cross the galaxy?

How little do we know of nature's processes and how much we assume. Truly, William Blake put it most succinctly: 'A fool sees not the same tree that a wise man sees.'

Perhaps then, since we know of long-term biological cycles such as these, there are other cycles hidden in nature, beyond the span of single lifetimes, that include the intrinsic ability of living creatures to adapt to environmental conditions presently absent from our planet. Indeed, does every creature, given the constraints of a changing environment – plus the biological loosening of periodic Sat yugas – have the capacity to change?

Just as the ermine changes it coat for winter; just as the seed can lie dormant for thousands of years; just as the bacteria and the rotifers can live in their desiccated time capsules for perhaps longer than we can ever envisage, awaiting a change of outer circumstances for the tiny living specks of dust to take on another form – just so, perhaps, may the living forms we know so well have secrets tucked away within them that only the rolling of the aeons can reveal.

The axolotl, the tadpole of the South American tiger salamander, is famous in the zoology textbooks for breaking with tradition and possessing the capacity to reproduce sexually while still retaining its aquatic larval form, including its external gills. But under little understood conditions of environmental change, its offspring can lose their gills, developing into the land-dwelling adults. In the lakes near Mexico City, at a height of 6000 to 7000 feet above sea level, it never ventures into adult form. Specimens up to a foot in length have even been recorded. Though in such a cycle, who is to say which is the adult and which the juvenile? In this form, it then remains, and when the rains return, it instinctively remembers it watery home and there lays its eggs, for the cycle to be repeated.

For many years, scientists thought the watery and land-dwelling forms to be *separate species*. It was only when some axolotls in captivity in the Jardin des Plantes in Paris bred, and their young lost their gills, becoming the well-known tiger salamander, that their secret was revealed. Now the change to the land-based form can even be induced in the laboratory by an energetic vibration playing the role of a key in the biological tapestry. It is a molecule we call a hormone.

Then there are the wet and dry season forms of some tropical butterflies which are thought to develop according to the temperature of the larva and pupa during their development.

In the case of *Melantis leda*, according to work done by Lt-Col. N. Manders, it is the temperature two to three weeks before the butterfly emerges from the pupa which regulates the form in which the handsome butterfly will emerge. This, effectively, constitutes a biological prediction of the rainfall and conditions with which the butterfly will have to contend during its adult life. An interesting piece of biological foresight, no doubt requiring many physiological modifications. And *how* we must wonder: how could such an intricate tapestry have evolved according to a linear process, involving only 'chance' as the driving factor?

So if species have a capacity for adaptation to small-scale environmental change built in, could there not be other potential changes hidden within the mosaic of form and function, both subtle and dense, we call a species? Changes which may not manifest even for millennia, until circumstances warrant it?

Digressing momentarily, it is interesting that those creatures which pass through various metamorphic stages, however different these outward forms may be, remain of essentially the same 'psychological' nature throughout their cycle. Given that the essential inward mental nature is present before birth, this is as one might expect. Thus a dragonfly and its larva are both voracious eaters of their fellow creatures. Butterflies and their caterpillars are mostly vegetarians, the one munching leaves, the other sipping nectar. And – most amazingly – even the seeds of the insectivorous pitcher plant – which traps its prey in pools of chemically laced water on leaves sporting a glissading escarpment of downward-pointing hairs and narcotic-laced nectar – even the tiny *seeds* of this plant trap prey by means of slime and poison, then digesting the surrounding mulch of decaying creatures as they germinate, forming into seedlings and plants.

In all cases, the inner mental being of a creature is similar throughout its life cycle, though expressing itself through different outward physical forms as it metamorphoses. But were creatures nothing more than their physical overcoats, why should they not be docile vegetarians in one stage and voracious prey-catchers in another?

Natural cycles there are in abundance, then. They can take many forms. And many, surely, that our human tunnel vision is unable to perceive. Ice ages, tropical ages, spiritualized eras, wet, dry, hot, cold, high, low – all these and many other uncharted cycles, our Earth must have undergone, and will undergo, over the millions of years of its existence.

The individual duration of one man's life spans just a fraction of a cosmic second in these dense, material vibrations. Then he, too, requires a brief respite from corporeal entombment. Perhaps the pause is only long enough for his mind to find and form a new, fresh and appropriate body in which to continue on the mental outworking of the karmic dance. Or maybe, the soul that has striven to rise above human weakness is given a brief, or longer, sojourn in some higher, astral sphere – until the effects of such aspirations are outweighed by the desires of the mind for more life experience in this denser, physical world.

Perhaps, too, the darker minds, full of rancour and disharmony, anger, lust, egotism, greed and intense attachment to the things of this world, takes the hell it experienced in physical life into its own world of mental hell when deprived of the physical form. For where else would a hellish mind go after death, than to a continued experience of its own hellish thoughts and dreams? But this, too is not forever, and then the soul takes physical birth once again. Though not, maybe in human form, but wherever the causal justice meted out by the Mind itself, may find its just fulfilment.

What does man know of these longer aeons? But one thing is sure. Man and creation are more than flesh and rock. There are more powers than chance and pure mechanism, underlying this great cosmic drama.

Even the surface of our planet is recycled in the most remarkable of ways. Its surface moves and folds, slips and slides, in a shifting patchwork of continental and oceanic slabs. The Himalayas, whose rise began maybe 50 million years ago, are still climbing heavenwards at an average rate of seven millimetres per year, double the speed of their advance ten million years ago, though such rates are by no means constant. Yet, running through them, the still more ancient river Indus had existed for millions of years before their rise. And cutting down through the rising rock, the Indus carries away from the Himalayan region five million tonnes of earthen debris every day. This mineralizes and fertilizes the flood plains below, as it slowly and inexorably finds its way to the sea. To these lands below, the Himalayas are a seemingly endless source of water, fertilizer and much else besides.

Over the last 1.5 million years, our planet has been in the grip of an ice age. Every 100,000 years or so there has been a 'temperate window' when the ice has receded for seven to ten thousand years. Then the ice has closed in again. In the temperate zones, lying just outside the ice-bound regions, when the minerals drain into the sea, the soil is depleted and less vegetation is supported. Then – says one of the theories – due to natural feedback processes affecting the climate, the ice again advances, the glaciers grinding the rock for remineralization of the topsoil in the next temperate window.

Night is required by plants to store the energy collected during the day. All creatures need sleep to replenish their energies for the next awakening. Winter is a necessary counterpart to summer, to concentrate and rest after the months of feverish activity. The dry seasons and the wet, the hot and the cold, all these are essential in the ebb and flow of nature's processes.

Just so, perhaps, are ice ages the long winters, the scourers of the planet, the deep sleep required for nature's recovery. Then life retreats to the equatorial zones, though there is little doubt that the ice-bound regions of such eras are far from devoid of life, any more than the Arctic and Antarctic zones are today.

But ice ages or no, millions of years of erosion will slowly flatten the planetary mineral heaps we call mountains. Not too fast – not too slow. The release is just right. But from where do new mountains, new planetary mineral repositories arise? If the land is all eroded into the sea in a matter of a few million years, how does the system keep on running? And what prevents the oceans from becoming so thick with minerals, like the Dead Sea, that no marine life can be supported?

The answer is to be found in the life forms themselves. Small sea organisms bind these marine minerals into their body structures, which then fall to the ocean floor upon their death. So continuous is this precipitation that it has earned itself the name of 'marine snow'. There the sediment accumulates over aeons, cleverly preventing the toxic over-mineralization of the sea, while at the same time providing the material for the rocks of future continents and mountains as the planetary surface gets recycled.

Man takes little account of such greater cycles. When his fields are starved of minerals through self-indulgent agriculture, he tries to impose his own will upon the land. But pouring fertilizers on to the fields does not help the situation, for the processes of nature are too complex to be so simply overridden or adjusted. Fertilizers are short-term exploiters, a reflection of man's mind, his greed for financial profit. For the plants man sows also depend upon microbes within the soil for their growth and existence. Microbes and bacteria play an integral part in the food cycle. Their activity is put to good account by plants, just as we and other animals eat the plants or even other animals and creatures.

Without bacteria, moulds and fungi, we would all be wading around in undecayed vegetation many miles deep. Not only that, but there are at least fifty known forms of bacteria which decompose rock into minerals. They have been particularly studied in the decay of old stone mansions, castles and other human monuments. Nature is so complete a system. These tiny creatures take the very rocks and stones as their food, releasing the valuable minerals into the soil upon which plant life may flourish.

But pesticides and fertilizers kill these microbes, turning fields into clinically sterile areas. The decaying mulch of good old manure and compost, full with a myriad bacteria, is required for a healthy soil ecosystem. The food we receive from plants grown upon these sterile patches is deficient in nutrient content and vibrational life energy. The degenerating planetary ecosystem becomes reflected in man's health. Degenerative diseases such as cancer and AIDS become rife. This is the natural process of nature's cyclic activity, the ebb and flow within time, space and the energy patterns of the creation.

To push a huge rock to the edge of a precipice requires tremendous effort and work. But once balanced there, it can be pushed over with just one finger. The energy required to move that system into a falling phase is very little. Once a system has been pushed too far out of equilibrium, it takes just a little further input for a period of turmoil to ensue, followed by a new equilibrium.

These threshold phenomena are an integral part of nature's processes. From the apparently indivisible quanta of physical energy, to the boundaries between the solid, liquid and gaseous states – the manyness of the energy patterns is prevented from becoming one homogeneous and ill-defined spectrum by threshold effects. Such clearly defined patterns are to be seen in both social and political, as well as all other systems.

Similarly, with our world ecosystem. We are poised on the edge. But it is not man's activities alone which can be held to account, for man is also a part of the natural ecosystem. Man is only responding to the greater cycles of nature.

So the mysteries of migration routes, which prompted this brief foray into the biological and geological past, is only one of a myriad miraculous facets of nature, of the greater Mind, that tells us of the great planetary drama in which life has existed, maintained within such finely balanced parameters, for hundreds of millions, if not billions of years.

This chapter ends with some brief comments on the nature of knowledge – scientific and mystic. Then in the next chapter, we tell some tales of the close kinship between man and beast, of contact between human minds and the minds of other living creatures. Finally, in the last two chapters, we talk at length of the greatest pattern maker of all, the power that weaves the one tapestry of nature, the greater or *Formative Mind*.

SCIENCE, KNOWLEDGE AND MYSTIC CERTAINTY

The insistence upon *rational objectivity* in science actually arises from an honest perception that we humans are capable of a high degree of self-deception. Lacking the certainty that arises from inner, mystic experience, we have a predilection to replace genuine knowledge with dogma. We therefore attempt to overcome this tendency by what we feel is the honesty of 'objectivity'.

At a human level, knowledge or science is supposed to be a pure objective description of experience, derived either directly from our senses or from instrumental extensions thereof. Such pure objectivity, however, does not exist, for our sensory or instrumental experience is meaningless without *interpretation*, which is carried out within our own minds. And our minds are heavily conditioned and circumscribed by habit, by what we have been taught to think, and by subconcious emotion, human weakness and social environment or circumstance. We see things within the context of what we are subconsciously programmed to see, through glasses tinted strongly with the colour of our inward, unconscious mind. This is as true of scientists as of any other group of humans, some would say more so. Even 'objective' experiments are devised according to certain thought patterns and beliefs.

Man's ideas do not chart a steady progression towards some ultimate philosophical or scientific truth. Mostly, they go round in circles, reflecting only the idiom of the day. In the context of history, no mundane ideas last for very long. It is only the basic spiritual truths which surface time and again, expressed through different idioms. Not so long ago, Western man believed that the Earth was flat, that the Earth was the centre of the universe, and that God had created the Earth in seven days according to a literal interpretation of a creation myth given in an old Jewish book. Scientifically, one can observe that the concepts of a geocentric universe and a flat Earth are fallacious, though there are still those who would disagree. Such observations would appear to be quite objective and true and I would agree. But these truths were by no means self-evident at the time of their first suggestion, when they were not so readily acceptable.

However, there are those theories that have their basis in a totally materialistic and mechanical *belief system*, not observation, and which interpret the evidence to prove the theory, ignoring or glossing over the anomalies. Such theories include the conventional model of evolution, Big-Bang models of the origin of material substance, the concept that life is a by-product of biochemical complexity, that mind and thought are to be found at the physical level of molecules and electromagnetism, and so on.

While human conceptual and even observational knowledge always carries with it an element of doubt, mystic knowledge is possessed of certainty, of direct perception and experience, quite undemonstrable to others. It is, however, immediately regarded by the fortunate experiencer as being of an altogether higher and more real nature than 'normal', everyday experience. Just as, when awakening from a dream, we automatically recognize its illusory reality and its creation from within the recesses of our own mind. You will find personal descriptions of such mystic experiences in my other books while an entire chapter in *Natural Creation: The Mystic Harmony*, is devoted to descriptions of mystic or semi-mystic experience.

Thankfully, this mystic certainty brings with it a deepening understanding of the ways of our human mind. This makes one quite comfortable with the fact that those who have never had such an experience will doubt its very existence and possibly even the sanity of those who experience it. One begins to see why people think and behave in the way they do. Increasingly, one sees the Mind underlying all forms. Yet the inherent authenticity and force of such mystical experience are the heritage of all and within the potential grasp of all human beings. Indeed, I believe that inklings of this understanding, 'Intimations of the Infinite', are experienced by many of us at least once in a lifetime and are available to all if we can simply open our hearts to the inward life which is calling us, but is obscured by the continuous dazzle and activity of our own superficial minds and senses.

Our lucid moments, those times when the shadow seems to fall from off the face of our inward confusion, these are the times when we are drawn closer to God, to our inward Source. They are the most real times of our entire life. Let no one, (not even yourself), ever convince you otherwise.

INTERLUDE

INTELLECT

The more we live by our intellect, the less we understand the meaning of life.

(William James)

Intellect is very good to live by in this world. We know the ways and means to spend our life comfortably with the help of intellect. But intellect does not lead us to the Lord. We have to leave the intellect in order to follow that path. What we require is practice and faith. I do not say an intellectual man cannot follow the path. Intellect is a barrier in our way, no doubt, but we have to pierce this barrier with the help of intellect. Unless the intellectual man is convinced that the path is right, he will never follow it. He can never have faith in the Lord as long as his intellect is not satisfied. But once his intellect is satisfied, he does not need intellect to follow that path. What we need then is faith and practice.

(Maharaj Charan Singh Ji, *The Master Answers*)

7. NATURAL COMMUNION

MAN, MIND AND CREATURE

Man can understand man. Or at least he has the potential to do so. Even if we meet with fellow humans from other parts of our globe where the spoken and written languages are different, the common, basic sense of meaning which we carry in our minds is enough for us to communicate through gesture and signs and to know that, should we care to, we can learn each other's language.

With other creatures, their communications are largely a matter of instinctive patterning. A dog never *learns* to go 'Woof', or a cow to go 'Moo'. They do so automatically even if raised on a bottle, by humans. It is as much a pattern of their inner being as is tail-wagging or chewing the cud. But whether man's inward mind patterning, which provides him with the capacity for verbal language, is underlain by a subtle and universal language of meaning to which we all relate when using verbal language, is an interesting possibility I discussed in *The Web of Life*.

Certainly, there is a basic mind structure in humans which we all recognize in each other while noting the absence of its full expression in other creatures. But in the Kal yuga, through which we now are passing, the age when mind predominates, man has largely lost consciousness of his inner centre of focused attention, the eye centre, and remains scattered in the world displayed to his mind by the five senses.

Man has thus lost contact with direct perception of his own inward mind energies and lives almost unconsciously, away from his true inner centre. Many of his natural mental capabilities such as telepathy and clairvoyance have therefore been submerged by the incessant activity of his lower nature, to the extent that folk even perform laboratory experiments to see if such faculties actually exist.

One might equally well undertake trials to see if we are really alive! But the real laboratory for all such experimentation is within our own body, in the inner areas of mind and soul. And the technique is called meditation.

Telepathy is not a sensory phenomenon. So to attempt to prove the existence of telepathy by the analysis of data acquired by sensory perception will always leave room for doubt. After all, there is no sensory evidence for the existence of thought, either. But do we doubt the existence of thought? Thought is a *mental* experience, not a *sensory* experience. We observe the *results* of thought everywhere, but not thought itself. Similarly with all aspects of mind – emotion, intuition, memory, intellect and so on.

So man communicates with man, but by word of mouth, rarely mind to mind. Thus we have suspicion, confusion and misunderstanding, for the mental meaning underlying words varies, as do our habituated beliefs.

And when it comes to other creatures, we are increasingly at a loss. Their inward minds possess different energy configurations to ours. Yet fellow members of a species can understand each other with the greatest of ease, while we may not even have the senses to perceive their communications to each other, let alone understand them.

We cannot see or smell pheromones, for example, in the way insects and other creatures can. We have no mental conception of what it is like to perceive the world through antennae. We are unable to hear the ultrasonic radar signals of bats, let alone interpret them with the intelligible rapidity which they can. We do not derive the same sense of meaning from a bee's waggle dance as its companions do. We hear the haunting sounds of the Humpback whales as they communicate under the sea, across vast distances, but understand little or nothing of what is passing between them.

Even with those creatures we assume to be our closest animal relatives, the chimpanzees and the great apes, we are forced into interpreting their signs and signals according to our own mental patterns. Such communications may clearly show signs of similarity to our own human allocation of emotions and feelings – anger, affection, dominance and so on – but of how a chimpanzee actually feels anger, we have little comprehension.

When, among humans, there are outbursts of anger or fighting, then such emotions normally leave a residue of resentment, of ill-feeling, which can take some time to subside or disperse. But Jane Goodall and others have reported that among chimpanzees, the enemies of the morning will be playmates of the afternoon. Very different from man's condition. No doubt this is all a part of nature's design to keep the community together, but of how a chimpanzee's inner mind is structured, and of how they feel, we have little notion.

Most humans, after all, have never successfully applied the scientific techniques of meditation to their own minds, including many of today's 'gurus'. So until we have fairly set out upon that road, how can we really hope to understand the inner beings of other creatures?

Their outward behaviour, *as observed through our mind and senses,* however – *not their's* – does result in some fascinating incidents and this chapter is devoted to the discussion of some of these. Other aspects were discussed in previous chapters.

WHALES, DOLPHINS, TIGERS AND DOGS

Many are the tales telling of the intelligence and concern shown by creatures of the animal kingdom. And there is no doubt that some of these animals live lives of great awareness, and of emotion, too.

When we see such intelligence and feeling among creatures which outwardly resemble ourselves – chimpanzees and other primates – we accept and can recognize it far more readily than we do among more alien creatures such as whales and dolphins. For apart from being equipped with eyes, head and body, dolphins and whales bear little outward physical resemblance to ourselves. Yet there is no doubt that they have an active, social life, full of real and caring communication, carried on in a language quite alien to our own experience of mind and meaning.

Many people, for example, will have seen the film of the Greenpeace anti-whaling efforts putting their little inflatable craft between the angry whalers and the hunted whales. And when they got into difficulties in the rolling ocean waves, the whales came to their rescue.

This may seem strange to us, but even among social insects, and certainly among birds and animals, there is a tremendous awareness of suffering, of danger to life and of mutual support in times of difficulty. Bees and wasps will attack en masse when their community is threatened. A bee sting emanates a powerful pheromone, perceived by other bees from a distance, which acts as an immediate and instinctive call to arms. Fish, reptiles, birds and mammals all protect their young and often each other, with great vigour, when danger threatens. Social species will frequently surround and protect a wounded or younger member of their clan, regardless of who the parents are.

Wherever there is social connection and association, then creatures respond to each other. Many are the cases of dogs, horses, bullocks, elephants and other creatures protecting the humans who are a part of their social group. Dogs, in particular, have been trained for this work – an extension of their natural instincts to protect the pack.

Jim Corbett tells just such a tale in his *Man-Eaters of Kumaon*, and since his lead-in to the episode also tells of the mental abilities of creatures other than man, it is worth quoting in full. In the story, Jim Corbett is describing his hunt for the Chowgarh tigress who killed at least 64 human beings between 1925 and 1930 before their first and last meeting took place. This, in fact, is a relatively small total compared to the hundreds which some man-eaters killed before they were themselves removed from the earthly scene. The villages mentioned are in the Kumaon region of the Himalayan foothills.

> On the road I had taken when coming to Dalkania there were several long stiff climbs up treeless hills, and when I mentioned the discomforts of this road to the villagers they had suggested that I should go back via Haira Khan. This route would necessitate only one climb to the ridge above the village, from where it was downhill all the way to Rainbagh, whence I could complete the journey to Naini Tal by car.
>
> I had warned my men overnight to be prepared for an early start, and a little before sunrise, leaving them to pack up and follow me, I said good-bye to my friends at Dalkania and started on the two-mile climb to the forest road on the ridge above. . . .
>
> The path wound in and out of deep ravines, through thick oak and pine forests and dense undergrowth. There had been no news of the tigress for a week. This absence of news made me all the more careful, and an hour after leaving camp I arrived without mishap at an open glade near the top of the hill, within a hundred yards of the forest road.
>
> The glade was pear-shaped, roughly a hundred yards long and fifty yards wide, with a stagnant pool of rain-water in the centre of it. Sambur and other game used this pool as a drinking-place and wallow and, curious to see the tracks round it, I left the path, which skirted the left-hand side of the glade and passed close under a cliff of rock which extended up to the road. As I approached the pool I saw the pug marks of the tigress in the soft earth at the edge of the water. She had approached the pool from the same direction as I had, and, evidently disturbed by me, had crossed the water and gone into the dense tree and scrub jungle on the right-hand side of the glade. A great chance lost, for had I kept as careful a lookout in front as I had behind I should have seen her before she saw me. However, though I had missed a chance, the advantages were now all on my side and distinctly in my favour.
>
> The tigress had seen me, or she would not have crossed the pool and hurried for shelter, as her tracks showed she had done. Having seen me

she had also seen that I was alone, and watching me from cover as she undoubtedly was, she would assume I was going to the pool to drink as she had done. My movements up to this had been quite natural, and if I could continue to make her think I was unaware of her presence, she would possibly give me a second chance. Stooping down and keeping a very sharp lookout from under my hat, I coughed several times, splashed the water about, and then, moving very slowly and gathering dry sticks on the way (as the local Indians might do), I went to the foot of the steep rock. Here I built a small fire, and putting my back to the rock lit a cigarette. By the time the cigarette had been smoked the fire had burnt out. I then lay down, and pillowing my head on my left arm placed the rifle on the ground with my finger on the trigger.

The rock above me was too steep for any animal to find foothold on. I had therefore only my front to guard, and as the heavy cover nowhere approached to within less than twenty yards of my position I was quite safe. I had all this time neither seen nor heard anything; nevertheless, I was convinced that the tigress was watching me. The rim of my hat, while effectually shading my eyes, did not obstruct my vision, and inch by inch I scanned every bit of the jungle within my range of view. There was not a breath of wind blowing, and not a leaf or blade of grass stirred. My men, whom I had instructed to keep close together and sing from the time they left the camp until they joined me on the forest road, were not due for an hour and a half, and during this time it was more than likely that the tigress would break cover and try to stalk or rush me.

There are occasions when time drags, and others when it flies. My left arm, on which my head was pillowed, had long since ceased to prick and had gone dead, but even so the singing of the men in the valley below reached me all too soon. The voices grew louder, and presently I caught sight of the men as they rounded a sharp bend. . . .

After my men had rested we climbed up to the road, and set off on what proved to be a very long twenty-mile march to the forest Rest House at Haira Khan. After going a couple of hundred yards over open ground, the road entered very thick forest, and here I made the men walk in front while I brought up the rear. We had gone about two miles in this order, when on turning a corner I saw a man sitting on the road, herding buffaloes. It was now time to call a halt for breakfast, so I asked the man where we could get water. He pointed down the hill straight in front of him, and said there was a spring down there from which his village, which was just round the shoulder of the hill, drew its water-supply. There was, however, no necessity for us to go down the hill for water, for if we continued a little further we should find a good spring on the road.

His village was at the upper end of the valley in which the woman of Lohali had been killed the previous week, and he told me that nothing had been heard of the man-eater since, and added that the animal was possibly now at the other end of the district. I disabused his mind on this point by telling him about the fresh pug marks I had seen at the pool, and advised him very strongly to collect his buffaloes and return to the village. His buffaloes, some ten in number, were straggling up towards the road and he said he would leave as soon as they had grazed up to where he was sitting. Handing him a cigarette I left him with a final warning. What occurred after I left was related to me by the men of the village, when I paid the district a second visit some months later.

When the man eventually got home that day he told the assembled villagers of our meeting, and my warning, and said that after he had watched me go round a bend in the road a hundred yards away he started to light the cigarette I had given him. A wind was blowing, and to protect the flame of the match he bent forward, and while in this position he was seized from behind by the right shoulder and pulled backwards. His first thought was of the party who had just left him, but unfortunately his cry for help was not heard by them. Help, however, was near at hand, for as soon as the buffaloes heard his cry, mingled with the growl of the tigress, they charged onto the road and drove the tigress off.

His shoulder and arm were broken, and with great difficulty he managed to climb on the back of one of his brave rescuers, and followed by the rest of the herd, reached his home. The villagers tied up his wounds as best they could and carried him thirty miles, non-stop, to the Haldwani hospital, where he died shortly after admission.

When Atropos, who snips the threads of life, misses one thread she cuts another, and we who do not know why one thread is missed and another cut call it Fate, Kismet or what we will.

For a month I had lived in an open tent, a hundred yards from the nearest human being, and from dawn to dusk had wandered through the jungles, and on several occasions had disguised myself as a woman and cut grass in places where no local inhabitant dared to go. During this period the man-eater had, quite possibly missed many opportunities of adding me to her bag and now, when making a final effort, she had quite by chance encountered this unfortunate man and claimed him as a victim.

By this time, the tigress must have come to know the particular scent, as well as stance, of her hunter and probably recognized him, even when dressed up as an Indian woman. As we have said, animals are superbly aware of the way a mind is reflected in body posture and there is no way that the hunter Jim Corbett could have concealed his alertness and readiness for attack, whatever disguise he adopted.

Then, too, the subtle contact of mind to mind, of life to life, is probably more a part of the tigress than the average man. Many hunters and men moving close to nature have commented on this sixth sense, warning them of danger.

Corbett also makes an interesting comment concerning tigers' sense of smell, which has a bearing on their natural instincts – man not being the natural prey of tigers.

> Tigers do not know that humans beings have no sense of smell, and when a tiger becomes a man-eater it treats a human being exactly as it treats wild animals, that is, it approaches its intended victims up-wind, or lies up in wait for them down-wind.
>
> The significance of this will be apparent when it is realized that, while the sportsman is trying to get a sight of the tiger, the tiger in all probability is trying to stalk the sportsman, or is lying up in wait for him. The contest, owing to the tiger's height, colouring, and ability to move without making a sound, would be very unequal were it not for the wind-factor operating in favour of the sportsman.

It is unlikely, of course, that a tiger actually thinks, as humans might – with our almost non-existent sense of smell – 'Which way is the wind blowing?' And then, sticking a wet finger in the air, 'I must get downwind.' It would seem more likely to simply be an automatic adjustment according to their immediate sensory experience and instinctive intent. Rather as one might almost unconsciously adjust the position of a book to get the best light from a bedside lamp without ever really thinking of angles and directions, so must the tiger get himself into the most appropriate location vis-a-vis the three-dimensional world of scents known to both himself and his normal prey.

Similarly, the fact that the tiger is unable to figure it out from our behaviour that we have practically no sense of smell, demonstrates how instinctive and preprogrammed are their mental functions.

On yet another attempt to kill the Chowgarh tigress, Corbett relates the following intriguing incident. The tigress had recently been seen in the area and Corbett was very much on the alert.

> For the next fourteen days I spent all the daylight hours either on the forest road, on which no-one but myself ever set foot, or in the jungle, and only twice during that period did I get near the tigress. On the first occasion I had been down to visit an isolated village, on the south face of Kala Agar ridge, that had been abandoned the previous year owing to the depredations of the man-eater, and on the way back had taken a cattle track that went over the ridge and down the far side to the forest road, when, approaching a pile of rocks, I suddenly felt there

was danger ahead. The distance from the ridge to the forest road was roughly three hundred yards. The track, after leaving the ridge, went steeply down for a few yards and then turned to the right and ran diagonally across the hill for a hundred yards; the pile of rocks was about midway on the right-hand side of this length of the track. Beyond the rocks a hairpin bend carried the track to the left, and a hundred yards farther on another sharp bend took it down to its junction with the forest road.

I had been along this track many times, and this was the first occasion on which I hesitated to pass the rocks. To avoid them I should either have had to go several hundred yards through dense undergrowth, or make a wide detour round and above them; the former would have subjected me to very great danger, and there was no time for the latter, for the sun was near setting and I had still two miles to go. So, whether I liked it or not, there was nothing for it but to face the rocks. The wind was blowing up the hill so I was able to ignore the thick cover on the left of the track and concentrate all my attention on the rocks to my right. A hundred feet would see me clear of the danger zone, and this distance I covered foot by foot, walking sideways with my face to the rocks and the rifle to my shoulder; a strange mode of progress, had there been any to see it.

Thirty yards beyond the rocks was an open glade, starting from the right-hand side of the track and extending up the hill for fifty or sixty yards, and screened from the rocks by a fringe of bushes. In this glade a kakar was grazing. I saw her before she saw me, and watched her out of the corner of my eye. On catching sight of me she threw up her head, and as I was not looking in her direction she stood stock still, as these animals have a habit of doing when they are under the impression that they have not been seen. On arrival at the hairpin bend I looked over my shoulder and saw that the kakar had lowered her head, and was once more cropping the grass.

I had walked a short distance along the track after passing the bend when the kakar went dashing up the hill, barking hysterically. In a few quick strides I was back at the bend, and was just in time to see a movement in the bushes on the lower side of the track. That the kakar had seen the tigress was quite evident, and the only place where she could have seen her was on the track. The movement I had seen might have been caused by the passage of a bird; on the other hand it might have been caused by the tigress; anyway, a little investigation was necessary before proceeding farther on my way.

A trickle of water seeping out from under the rocks had damped the red clay of which the track was composed, making an ideal surface for the impression of tracks. In this damp clay I had left footprints, and over these footprints I now found the splayed-out pug marks of the tigress where she had jumped down from the rocks and followed me, until the kakar had seen her and given its alarm-call, whereon the tigress had left the track and entered the bushes where I had seen the movement. The tigress was undoubtedly familiar with every foot of the ground, and not having had an opportunity of killing me at the rocks – and her chance of bagging me at the first hairpin bend having been spoilt by the kakar – she was probably now making her way through the dense undergrowth to try to intercept me at the second bend.

Further progress along the track was now not advisable, so I followed the kakar up the glade, and turning to the left worked my way down, over open ground, to the forest road below. Had there been sufficient daylight I believe I could, that evening, have turned the tables on the tigress, for the conditions, after she left the shelter of the rocks, were all in my favour. I knew the ground as well as she did, and while she had no reason to suspect my intentions towards her, I had the advantage of knowing, very clearly, her intentions towards me. However, though the conditions were in my favour, I was unable to take advantage of them owing to the lateness of the evening.

I have made mention elsewhere of the sense that warns us of impending danger, and will not labour the subject further beyond stating that this sense is a very real one and that I do not know, and therefore cannot explain, what brings it into operation. On this occasion I had neither heard nor seen the tigress, nor had I received any indication from bird or beast of her presence, and yet I knew, without any shadow of doubt, that she was lying up for me among the rocks. I had been out for many hours that day and had covered many miles of jungle with unflagging caution, but without one moment's unease, and then, on cresting the ridge, and coming in sight of the rocks, I knew they held danger for me, and this knowledge was confirmed a few minutes later by the kakar's warning call to the jungle folk, and by my finding the man-eater's pug marks superimposed on my footprints.

What all this indicates is that creatures – human or otherwise – are aware, are conscious, of danger to life and limb. And that all creatures have a mind. The behaviour of the buffaloes, the kakar, of Jim Corbett himself, and the hunting plans of the tigress all tell us this in no uncertain terms. We see this all the time in the behaviour of the life forms in our gardens. The sparrows fly away when a marauding pussy appears on the scene. They mentally appreciate the arrival of danger and act accordingly. There is a mental process underlying all actions in all creatures, just as there is with man.

Yet, somehow, because the mind of another creature is such alien territory to us, we find it difficult to accept that such mental activity and subjective sensory *awareness*, really is going on.

It is intriguing, of course, to speculate upon the intelligence of creatures so apparently alien to ourselves as tigers, whales and

dolphins. Yet familiarity may be blinding us to equal intelligence
expressed by animals far closer to home. A collie dog I know, who
lives with my neighbours, has an old semi-deflated toy football which
she never tires of carrying to the feet of any human who will kick
it, for her to gallop off, retrieving it and placing it once again near
one's feet. She stands off expectantly to one side, limbs and body
ready for an instant gallop or spring, her eyes focused upon you in
an attempt to induce an endless repetition of the game. There is no
doubt of her mental motivation or desire to simply play and exercise,
to express what she feels in her mind is the most natural of things
to do.

We get quite used to our domestic dog actually bringing us sticks
and balls so that we may throw them for our four-legged companion
to retrieve. We are not amazed at their motivation and intelligence
when they eagerly look up at us, willing us to throw the ball or the
stick. And there is no doubt of their thorough enjoyment of the game.
We even join in the playful antics of our puppies and kittens. But our
curiosity is aroused when we see such alien creatures as dolphins and
whales going through similar routines.

Our dogs are affectionate, interested, playful, forgiving, loyal,
intelligent, spontaneous, creative and much more besides. They
even have a language, so most definitely they possess a mind
structure. And so, too, do all creatures. This is where behaviour,
instinct, cognition, awareness, sensory experience, memory, learning,
all their other faculties are lodged. How could we ever have believed
otherwise?

THE CHIMP AND THE GORILLAS

Reference to chimpanzees and apes has been made here and there
throughout the preceding chapters and it seems appropriate to relate
just one example of their mental processes in these pages. This is taken
from Sue Savage-Rumbaugh's book, *Ape Language.*

Sherman and Austin were two chimpanzees who had been trained
to use a computer keyboard in which the keys carried lexigrams
(geometric patterns) which related to items (e.g. JUICE, SWEET POTATO)
or simple activities (e.g. GIVE, GET, CHASE).

One day the following incident took place.

Two teachers were working with Sherman and Austin on locations. In
the course of this training, one teacher would go to a specific location
(sink, outdoors, playroom, etc.) while the second teacher remained with

the chimpanzees by the keyboard. The chimpanzees were asked to label the location of the teacher who was moving about.

During one of these trials, Sherman refused to label the teacher's location, saying 'Collar' instead. He then walked out of the room and directly to the area where the collars the chimps had to wear outdoors are usually kept hanging on the wall. 'Collar' was thus often used as a way of asking to go outdoors, and after saying 'Collar', the chimpanzees would often go search for them and put them around their necks while pant-hooting in anticipation.

On this particular occasion, Sherman saw no collar hanging in the usual spot, and so, after looking carefully, he then walked to the toy shelf (where the collars are sometimes inadvertently left) and searched through the toy bins. After a one-minute search, he noticed the *National Geographic* magazine which he had been looking at with the teacher earlier in the day.

Koko, the signing gorilla, was the main feature in this issue. When Sherman found this issue on the shelf, he tapped the pictures of Koko with his hair standing on end (pilo-erection) and swaying back and forth bipedally. He then picked up the magazine, spied his collar on the floor and returned to the teacher.

He held out the collar to the teacher with one hand and pointed to a picture of Koko with the other. When the teacher said 'Yes, go outdoors,' Sherman repeatedly uttered high-pitched barks of excitement and started running around and around the lab with the magazine.

Going out-of-door was, of course, the most favoured activity of Sherman and Austin and it was not unusual for Sherman to be excited about going outside. What was unusual in this case was that even in his great excitement he still carried the photographs of Koko and pointed to them repeatedly as he looked toward the teacher, who nodded and said 'Yes' to Sherman.

The teacher knew that Sherman was combining the presentation of the collar and Koko's picture to communicate something rather specific but was uncertain as to what. The teacher also knew that when she answered Sherman's request in the affirmative he appeared to be extraordinarily pleased. As the teacher put the collar on Sherman he continued to point at the picture and bark very loudly.

Outdoors, he continued to carry the picture while walking bipedally, and he headed directly for the gorilla quarters which were located near the adult-ape house on a hill behind the language lab.

On a few previous late evening occasions Sherman had been allowed to go to this area and view the gorillas, an event which both scared and thrilled him. However, in general, he was not allowed to go there since his presence excited all of the large apes housed there and interfered with any research that might have been going on. Generally, if Sherman tried to lead his teacher towards this area, she would refuse to allow him to go.

The closer Sherman got to the gorilla housing area, the faster he ran, the louder he barked, and the tighter he held the pictures of Koko. Thus

Sherman's intent and expectancy became clear to his teachers. From his point of view he had asked to go to see the gorillas and he had been told 'Yes'. However, the teacher had not initially understood the request completely and the acquiescence with regard to going outdoors had led Sherman to believe that he could also go see the gorillas.

When it became apparent to the teacher that Sherman was intent on proceeding to the gorilla area, it became necessary to say 'No.' As Sherman saw the teacher stop and point in the other direction, he dropped to a quadrupedal stance, put down the *National Geographic*, stopped vocalizing completely, lost his pilo-erection, and appeared completely dejected. He showed no further interest in the magazine or in being outside.

Although Sherman had not used the keyboard, he had certainly made a combinatorial request. It had involved the use of a symbol and a photograph to produce a novel communicative request to go to a very specific location, and one that he could not request at his keyboard alone. The marked change in Sherman's demeanour when he realized that his request was not going to be granted was dramatic. His immediate discarding of the magazine at this point further supports the view that he had indeed been intentionally using the magazine to communicate his desire to go and see the live gorillas.

It is important to point out that Sherman and Austin had never been taught to use photographs as a means of requesting to go to locations. Moreover, this particular photograph had been brought to the lab only that day, and no one had ever coupled it with any sort of training.

Within the context of our discussion, this incident speaks for itself.

MAN MEETS BEAST

Many naturalists and others are well aware that the content of one's own mind determines the way in which not only members of our own species react to us but also those of other species, too. Let me quote, here, from Allen Boone's delightful book, *Kinship With All Life*. Boone's inspiration came from his personal experiences as temporary 'keeper' of the amazingly intelligent German shepherd dog, Strongheart, who became a Hollywood 'film star' back in the 1950's. It was Strongheart's manifest mental activity which opened his eyes to the mental processes present in all living creatures.

Rattlesnakes

There is great practical value in the art of carefully supervising one's thoughts and motives in contacts with other living things. Particularly with such creatures as rattlesnakes. These wise but little understood fellows, with their poison-brewing skill and deadly defense techniques, are experts of the first magnitude in dealing with thought emanations, especially as they come from a human.

When I first visited those parts of the West where white men and Indians frequently crossed trails with rattlesnakes, and when I had to do so myself, it was a shivery experience. I saw some of them whirl into action with their hypnotic eyes and their lightning lunges with heavily poisoned fangs. They were thorough, terrifying and deadly killers.

One day an old desert prospector, who had had rattlesnakes as neighbors for as long as he could remember, told me a surprising thing. He said that while rattlesnakes take special delight in sinking their fangs into a white man, they seldom harm an Indian. I asked him why. He did not know, and had never tried to find out.

In my travels I found that what the old prospector had said was true. The rattlesnakes were indeed selective. They were biting the white men, and they were extending almost complete immunity to the Indians. I talked to all kinds of 'snake experts', but none of them gave me a satisfying answer; certainly none gave me an answer that I would have wanted to try out on a diamond-back rattler.

Almost everywhere I went there was vicious and relentless warfare going on between white men and rattlesnakes; it was warfare to the death of either the man or snake. But I could find no such warfare between the Indians and the rattlesnakes. There seemed to be a kind of gentlemen's agreement between them. In all my journeyings in deserts, prairies and mountains I never once saw a rattlesnake coil, either by way of defense or attack, when an Indian walked into its close vicinity.

My dog-trains-man sessions with Strongheart had shown me the trouble that unseen mental forces can cause in one's contacts with animals. And so I was able to understand why there was warfare between white men and rattlesnakes, but practically none at all between the Indians and the snakes. This situation between the humans and the snakes confirmed what Strongheart had been so patiently trying to teach me: that one's thinking, in all its nakedness, always precedes him and accurately proclaims his real nature and intention. . . .

The answer had to do with individual states of mind, with the kind of character atmosphere that was being diffused, with projected thought-things or forces. Almost every rattlesnake that I watched illustrated this illuminating relationship point for me. The snakes were able to detect and correctly appraise the particular kind of thinking that was moving in their direction. Having done so, they were ready to deal either as a friend or as foe with the approaching human body belonging to that thinking.

What really happens when the average white man and a rattlesnake suddenly and unexpected meet? Having been taught to regard all snakes as loathsome and deadly enemies with no rights whatsoever on earth, the man wants to kill every snake he sees. Something intensely emotional. It is a kind of thought-vendetta, a condition of mutual ill-feeling in which each strikes at the other with destructive attitudes and intentions.

If the white man happens to have a material weapon and is able successfully to use it, he kills the snake's physical body. If, however, the snake manages to avoid the blow and gets within range, it buries its well-poisoned fangs in some part of the white man's body, and the man keeps the rendezvous with death. While the snake may victoriously jab its fangs into the white man's body, what it really strikes at is the unsocial and deadly thinking that animates the body.

Watching a real American Indian walk into the vicinity of this same rattlesnake, you witness something entirely different. For one thing you would be unable to detect the least sign of fear or hostility in either one. As they came fairly close, you would see them pause, calmly contemplate each other for a few minutes in the friendliest fashion, then move on their respective ways again, each attending strictly to his own business and extending the same privilege to the other. During that pause between them they were in understanding communication with each other, like a big and small ship at sea exchanging friendly messages.

Could you look deep into the thinking and motives of the Indian, you would discover the simple secret of it all, for you would find that he was moving as best he knew how in conscious rhythm with what he reverently calls The Big Holy, the great primary Principle of all life, which creates and animates all things and speaks wisdom through each one of them all the time. Because of this universally operating Law, the Indian was in silent and friendly communion with the big rattler not as 'a snake' that had to be feared and destroyed, but as a much-admired and much-loved 'younger brother' who was entitled to as much life, liberty, happiness, respect and consideration as he hoped to enjoy himself. His 'younger brother' had reacted accordingly.

It is not only man and animal who are closely connected, mentally, at a subtle level, but all creatures, including man and man, or man and woman. When folk harbour ill feeling, intolerance, criticism or prejudice in their minds against anyone, these vibrations emanate from them, impinging – mostly unconsciously – upon the minds of those against whom they feel mentally at odds. This happens even in the absence of their *bête-noir*, but is more concentrated whenever there is direct personal contact. Then the sparks can fly, and if neither party is really aware of the power of their own minds as the true architect of such situations, anger and emotion will rule the day.

Thus, in a family situation, for example, if one member entertains negative and unfriendly attitudes towards another, then whenever they meet there is likely to be trouble. The *apparent* cause of an argument may be quite trivial, but the deeper cause lies in the often unconscious attitudes of one or both parties.

Fortunately, the reverse is also true. For the one who encourages nothing but genuine feelings of peace, love and goodwill to all his fellow creatures, generally finds that peace and harmony follow him wherever he goes. Love brings out the very best in both ourselves and others. 'Everyone is a very nice person all the time, if we only adopt the most loving and positive of attitudes towards all our fellow human beings,' as a great man once said.

However, returning to Allen Boone's story of the rattlesnakes:

Tail-Rattlings

Even the most dreaded of poisonous snakes is a kindly disposed fellow at heart . . . This is no doubt hard to believe, but I have seen it proved again and again in various parts of the world with all sorts of 'killer' snakes. The proof has been established by a number of unusual men and women, all of whom moved out from the same operating basis . . . the premise that harmonious relationships are possible only after they have first been made so mentally.

One of the most interesting of these rare people was a slight and most unassuming little woman named Grace Wiley, who until the summer of 1948 made illuminating snake history at her Zoo for Happiness not far from Long Beach, California. Miss Wiley had long experience as a herpetologist and was considered one of the world's most skilled handlers of snakes with bad reputations. Indeed, the tougher, the meaner and the more venomous they were, the better she liked them. In her Zoo for Happiness one could find almost every known kind of a snake – enormous king cobras over twenty-five feet long, Egyptian cobras, adders, copperheads, vipers, Australian black snakes, green mambas, tiger snakes, fer-de-lances, moccasins, all varieties of rattlesnakes and many others.

People came from all over to see the collection and to watch Miss Wiely handle the snakes. Visitors were permitted to handle them, too, under her watchful supervision. People also were drawn there as students to listen to her interpret the snakes from the point of view of the snakes themselves. With one of the deadly specimens nestling affectionately in her arms, she would show to fascinated audiences what splendid . . . companions snakes can really be when given an opportunity. She would usually close these talks with the observation that deep within its heart the snake is not a troublemaker but a fine gentleman, and that when he strikes he does so because someone with evil intent has invaded his domain and cornered, frightened or hurt him.

Watching this soft-spoken little woman prove these principles with all kinds of dangerous snakes was a breath-taking experience. This phase of her work was carried on in what was known as the 'gentling room', a severely bare place with a heavily built, oblong table in the exact center. Most visitors were not permitted in this room when she was gentling a snake because of the danger involved, but there was a glass arrangement in one of the doors through which the privileged few were able to watch what went on.

Watching from this safe point of vantage, one sees Miss Wiley quietly enter the room, take a position just off the far end of the table and become as motionless as the table itself. In each hand she holds an odd-looking stick about three feet long. One of these has a cuplike mesh arrangement on the end of it; this is used for stopping and pushing back the heads of striking snakes. The end of the other stick is padded with soft cloth and is known as a petting stick.

A large box with warning signs all over it is wheeled into the room and placed on the table. Loud, clattering, spine-tingling sounds coming through top and sides of the box proclaim the presence of a rattlesnake. At a nod from Miss Wiley the rear end of the box is elevated, the front part jerked off, and out into a new world of experience slides Mr Snake. And what a snake! He is over six feet in length, beautifully designed, filled with tremendous energy and just about as deadly and menacing as they come. He is a magnificent specimen of diamond-back rattler, newly arrived from deep in the heart of Texas, where snakes grow plenty big and plenty tough.

As the snake hits the table there is a flash of movement almost too fast for the eyes to follow, the swift coiling of its body into a defense or attack position. The big fellow from Texas is set to fight anyone or anything for survival. But to his obvious astonishment and bewilderment, there is nothing to fight. There is no moving target to strike. Only the bare walls and the motionless woman facing him. The snake's head darts apprehensively in all directions trying to discover from which direction trouble is going to come. His tail rattles furious warnings. But nothing happens. Nothing at all.

Why does Miss Wiley not do something with the sticks in her hands? Why does the snake with all its noise and threatenings, not take at least a practice lunge at Miss Wiley with its poison fangs?

The truth is that Miss Wiley had been doing a most important 'something' to the big snake every since it came sliding out of the box, but you could not tell she was doing it because it was entirely mental. What was really happening was not just the outward meeting of a woman and a snake; rather it was the exploratory coming together for the first time of two invisible individualities . . . of two states of mind . . . of two puzzled and wondering kinsfolk who were about to discover that they really are related in the great Plan and Purpose of Life.

From the moment that Miss Wiley first saw the big snake, she had been silently talking across to it. Outwardly she appeared to be doing nothing at all. Actually she was proving the potency and effectiveness of her favourite rule of action in all relationship contacts: that all life, regardless of its form, classification or reputation, will respond to genuine interest . . . respect . . . appreciation . . . admiration . . . affection . . . gentleness . . . courtesy . . . good manners. The big tail rattler was being lovingly showered with these qualities, undoubtedly for the first time in its experience.

Had your ears been attuned to the silent universal language of the heart, you would have heard in detail the flow of soundless good talk that was moving from Miss Wiley to the snake. Not down at it as 'a lower form of life', but across to it as a fellow expression of life. And in that good talk, among other things, you would have heard her praising the snake for its many excellent qualities, assuring it that it had absolutely nothing to fear, and reminding it again and again that it had simply come to a new home where it would always be appreciated, loved and cared for. All of this was communicated without the slightest sound or gesture from Miss Wiley.

After a while you would notice a marked change in the snake's attitude. The fast rattling of its tail was slowing down. Its head, which had been glaring in all directions at a fast, nervous tempo, was steadying itself in the direction of Miss Wiley, even though it could not clearly distinguish the motionless woman from the motionless wall behind her. The 'killer' from Texas was not only feeling but actually responding to the friendly thoughts and feelings being sent in its direction.

As Miss Wiley continued her reassuring talk, but now in low, soft vocal tones, you would witness the blossoming of this unique gentling technique. You would see the big snake slowly uncoil and cautiously stretch itself the full length of the table, finally resting its head within inches of where Miss Wiley was standing. Then the first physical movement by Miss Wiley as she reached across and began gently stroking the snake's back, in the beginning with the soft-padded petting stick, and then, there being no resistance, with her two bare hands. As you watched this almost unbelievable performance, you would have seen the snake arch its long back in catlike undulations, in order better to feel the affection-filled ministrations.

At that precise moment in the gentling process another 'deadly poisonous snake' had become a member in good standing of the Zoo for Happiness. Miss Wiley had again demonstrated the fact that regardless of appearances, good is latent in every living thing, and simply needs to be called into active expression through the gracious application of respect, sympathetic understanding, gentleness and love.

This excerpt really says a great deal. Whether man or beast, we are all of the same essence, we all have a soul. Rather, we *are* souls. And all souls are one in the great ocean of the Supreme Consciousness, the Lord. Soul is the same as consciousness or Life. The same essence of life is within all creatures. What names you give it are irrelevant.

Around that soul, we have our mind. This separates human from human, and creature from creature. But all the same, these minds are drawn from the great ocean of the Universal Mind. They are of a similar subtle energy or material, just as our own bodies are. So when, in our mind, we feel kindly disposed to our fellow creatures, human or otherwise, then they also feel kindly disposed towards us. The subtle vibrations of our mind reaches out and touches the minds of all others.

In this way, too, the angry man draws the discord he subconsciously craves down upon his own head. The lustful person puts out strong sexual energy and attracts the same, while the peaceful and loving heart finds himself automatically with those of a similar disposition. We both find and create in others that which we consciously or unconsciously have in ourselves.

There are many tales of naturalists who have gone to some place in search of a rare species, only to find that a member of that very species floats down out of the trees on gentle wings to besport itself before his amazed eyes, or appear in whatever appropriate manner to his appreciative gaze.

This was the experience of the chaplain from my old school, a man from whom I learnt many secrets of the natural world on hiking trips through the mountains and lakes, sea shores and high cliffs of the English Lake District. His name was Mr Moule and, being a reverend, was promptly christened, 'Holy Moly' by all the children. He found many rare species of bird, plant and insect upon his rambles, and understanding man's destructive nature, he told very few folk the location of his finds.

He even came across a restricted locality where lived a colony of what must have been a new sub-species of the extremely rare, Small Mountain Ringlet butterfly, for they were at a much reduced altitude to their normal haunts, and altogether smaller than the regular species. Knowing of my interests, he showed me the location, but preferred the continued existence of the species to the mild fame which would have accompanied his reporting the discovery in some scientific journal. And though I have not been there at the right season for many years, for all I know they live there still – a small, brown, undemonstrative little butterfly, not immediately noticed by an untrained eye.

On another occasion, Holy Moly set off in his Morris Minor hoping to see the rare Purple Emperor butterfly in some southern oak wood. This magnificent insect, as well as being rare, frequents the tops of oak trees, rarely descending to the forest glades, so binoculars are normally required to catch a sight of it. It does, however, have a penchant for glistening muddy puddles and shiny motor cars. But not every car.

But now, with the car hardly parked and before the good reverend could open the door, down from the trees swooped a resplendent male Purple Emperor. Alighting upon the bonnet (hood, if you are American), in full view through the windscreen, it opened its wings to their fullest extent, remaining there before the grateful gaze of this gentle man who walked with God and loved the creatures of His natural world.

So the naturalist, walking with a peaceful heart and a quiet mind sees many things. Creatures often seem to have gone out of their way to find him and display themselves to him. Yet the hunter, walking with gun on arm, thoughts only of killing and murder in his mind and his intent clearly expressed in his body language, is shunned in fear by all creatures who can. This makes good sense, you will agree.

PLANTS THAT TALK

Nature may be beautiful but there is little doubt that Darwin was right when he observed that only the fittest survive. There is much cooperation, co-existence and mutual dependence – but there is also dynamic and ruthless competition. Nature's battles rarely erupt into war, yet they contain some of the most intricate and complex of defensive and predatorial designs. For life lives upon life: what is life to one is dinner to another! Even plants know how to attack other creatures and defend themselves.

Many insects and their larvae eat the leaves of plants, and in recent years a complex skein of plant-insect interactions have been glimpsed. Some insects, for example, emit a pheromone as an alarm signal, warning other members of the same species of impending danger. But we also find that certain plants cleverly emit the same pheromone, and the insect flies away without molesting it.

Pheromones are complex molecules requiring a continuum of energetic interchange through many 'stages', for their formation. For insects to have *evolved*, by chance, not only the ability to *emit* these chemicals on sight of danger, but also simultaneously to interpret the chemical as a messenger of alarm, would seem to be an impossibility.

The biological and physiological complexity required for molecular recognition either 'directly' through the senses of 'smell' or 'taste', or 'indirectly' through appreciation of their modulating effects upon electromagnetic energy, is quite amazing. The antennae and associated cellular structures are 100 per cent accurate and yet amazingly intricate. It requires a far finer tuning than that of any ordinary human radio to be able to pick up specific molecular electromagnetic emissions, or to sense a pheromone in any other way. The complex shape of the aerial – the antennae – has to be perfect. And then, the myriad interactions in the biochemistry and physiology of nerve and brain must somehow be integrated to perfection. Surely, such a thing could never have evolved in tiny steps by chance mutation?

Moreover, such 'evolution' by chance mutation has no goal in mind. So if natural selection of random mutations were truly the driving factor underlying change among the creatures, why is nature not riddled with biological imperfections at a biochemical and physiological level? Why are there no half-formed sense organs or biochemical networks, with a few links still missing? However, this is a subject we return to with full attention in *Natural Creation . . . Or Natural Selection?*

So plants, quite naturally, have the ability to protect themselves from predators, just as all other creatures do. Some emit poisonous gases which kill or temporarily paralyze any insects who attempt a meal. This gaseous emission may be constant, but frequently it comes entirely as a response to a predatorial nibble.

But while this method keeps many plant-eaters at bay, there are always a few creatures who have the design capability to deal with the toxins. No creature in this world is left without some enemies. This is a natural part of the outworking of the karmic law of the mind.

The common European weed, ragwort, for example, contains various alkaloids – poisonous to most insects and animals alike. Yet the red and black cinnabar moth has caterpillars which, while they are happy to eat innocuous weeds such as groundsel, are also ragwort specialists.

These clever little yellow and black-banded caterpillars actually store the alkaloids in their own bodies, giving them a bad and poisonous taste to any would-be predator. And not only that, but they pass on their chemical deterrent through the pupal stage into the adult moth itself. The bright, but differing, colours of the moths and caterpillars are warning signs, too, easily recognized by any insectivorous creature in search of a meal, and readily learnt should the first warning not be enough.

Again, how could the ability to eat a 'previously' toxic plant, ever be *evolved*? The biochemistry required to deal with alkaloids is not a matter controlled by one or even two gene patterns in the DNA. It is a vastly complex process and unless you get it right in one, you will never live to tell the tale, to get it right the next time, in order to 'evolve' the skill! So how could a species slowly and randomly evolve the ability not to be killed outright?

And so the story continues. Every creature has its own tale to tell. The leaves of cabbages, mustard and some other brassicas contain toxic mustard oil, similar to the deadly mustard gas. But along comes a cabbage root fly, whose larvae feed on the roots. The female fly lands on the plant and tests it for 'taste', first with 'taste-buds' on her feet and then with her proboscis. If the plant checks out successfully, she jumps lightly to the ground and lays her eggs – not in the roots – but in the soil. And when the minute grubs hatch, they are immediately attracted, through some as yet unidentified sense organs, to the scent of mustard oil emanating from the roots. And by untaught instinct they know that that means food. One creature's poison is another's meat.

Then there is the common bracken, a fern which covers the hillsides in many areas of Europe. The sheep wisely leave it alone and most bugs do, too, for it contains two chemicals which, when a mixed together in a creature's stomach, make cyanide. A deadly cocktail to be avoided by most organisms. But not so the saw-fly larva and certain other specialist feeders, who possess the biochemistry to handle the toxic diet.

Indeed, bracken also contains other unpleasant substances, terpenes, which, like the cinnabars and alkaloids, are harvested by the saw-fly larvae and stored in their bodies, making them unpalatable to predators. Yet saw-fly larvae are actually pretty omnivorous. So one wonders why they should have the ability to eat such a normally poisonous diet when there are so many other foods they can eat. And how could such a specialist, even unnecessary, sideline have ever evolved?

But perhaps the *pièce de résistance* are the trees that 'talk'. It has been observed that the individual leaves of many trees, when damaged, increase their internal quantity of various tannins, thus making themselves unpalatable to further would-be nibblers – whether insects, cattle, deer or other species.

The effect, however, lasts only for about twenty-four hours, for tannin manufacture (say the botanists) consumes a considerable amount of plant energy and resources. And there is no need for a creature to keep on running when the chase is over. But the onset of tannin production is very rapid. Sometimes within just a few minutes of the arrival of leaf-eating creatures, the tanin content is up to levels which make not only the leaves under attack, *but all the other leaves of the tree* toxic or at least unpalatable. This limits the degree of possible grazing or nibbling and explains why deer and cattle, for example, will eat just a few mouthfuls from each tree before moving on. A co-operative arrangement which keeps the trees healthy and prevents the animals from killing off their food supply.

The immediate assumption is that there is a chemical communication between leaf and branch all over the tree, and this possibility is borne out to some degree by the character of leaf and caterpillar battles in other species. There the caterpillar first cuts all the leaf veins leading back to the main stalk before settling in for a dinner which it hopes cannot then be communicated to the neighbouring leaves.

But this cannot be the whole story, for David Rhodes and others have discovered that *this communication can even take place from tree to tree*, where there is no root or other physical contact. Now it is known that ripening tomatoes ripen faster when placed together on a window ledge, because they produce ethylene gas which stimulates their neighbours to faster and synchronous ripening. Do plants, then, also communicate by means of airborne pheromones or in some other, perhaps more subtle fashion, as well? This is an amazing scenario. It means that similar – perhaps even dissimilar – species of plants and trees, growing in a group are in constant communication with each other. It makes one look again at the plants in one's garden and wonder what secret messages are being passed in ways too subtle for our human senses to grasp.

For if they are able to communicate danger (what greater danger to a plant can there be than something destroying its leaves or roots?), one wonders what other signals are they able to pass between themselves? The phenomenon of companion planting – growing different species near each other because they thrive better – is well known, though there is frequently no known reason why this should be so.

I have also noticed many times in my garden, how some strong-growing 'weed' appears to 'intimidate' those plants who were previously growing there. They recoil away as if the intruder were unpleasant to them – something more than simply shading them from sunlight. Yet plants often thrive when the same species are massed together. Then they do not recoil from each other, but blend together in one mass. So do such 'predators' of the plant kingdom emit unpleasant plant pheromones or some other disharmonious vibration, some subtle energy or perhaps some electromagnetic (e.g. infrared or microwave) oscillation, to disturb and 'attack' the plants whose space they wish to overrun?

After all, their species-specific infrared radiations are detectable by insects – either as 'visible scents' from flower and leaf and root, or from the leaf and hair structures behaving as frequency-specific aerials or from other biophysical processes which we do not understand. So why should the plants themselves not communicate with each other by these same means?

Trees and plants do project different subtle atmospheres around them, as Dr Bach discovered when he used some of their flowers to produce remedies contaning their subtle vibration. If humans can feel this, then why not the trees themselves? The feeling under an old yew tree is quite different to that which one experiences beneath a chestnut, a willow or an oak. These are the subtle mind currents which emanate from all living creatures and constitute a part of their aura. If you have never experienced this, try it out for yourself one day when your mind is calm and you are not with companions who will ridicule you or otherwise disturb your peace of mind. Reasonable inner quietness is essential for awareness of the subtle.

Every soul takes birth in a form appropriate to its mental qualities – its desires, impressions and entanglements from previous lives. These are expressed in the outer form and 'life-style' of the creature – whether plant, tree, insect, animal or man. The mental vibrations are also expressed in the aura which every incarnate being has around it. In fact, it is this mental or auric vibration which shamans and herbalists feel when deciding which herbs are appropriate for which people and which condition. For the mental vibration is directly expressed in the biochemistry and 'life-style' of the plant. By observing the whole plant – gross and subtle – by listening with their mind as well as perceiving with their sense, nature's secrets can become known without the need of costly research laboratories.

Nature is perfectly balanced. Everything that happens is by design and purpose. Nothing is out of place. From the wider point of view, this even includes man's disharmonious activities, for nature knows how to keep her own balance and even the disharmonies of a century are a very small matter in the context of the ages. No species is inviolate, else the individuals would live indefinitely – a soul forever entombed in the same body; and this is not the purpose and meaning of life. Life lives on life – or rather physical bodies live on physical bodies – in a closely woven tapestry of creation, preservation and destruction. But the harmony of the whole is always maintained, though there can be changes of emphasis. But just as the overall form of our body remains the same, despite the constant replacement of the atoms and molecules of which it is made, so is the overall form of nature maintained as identifiably the same old physical universe, though the pattern may alter markedly over the course of time.

But the real source of interplay is at the inner, formative levels of Mind and life. And that is the subject of the remaining chapters.

INTERLUDE

INTELLIGENT LIFE ON EARTH?

For many months, the archway above the entrance to Cambridge University's Department of Applied Mathematics and Theoretical Physics carried the graffiti: 'Is there intelligent life on earth?'

The query must have been considered one of some value, as well as entertainment: for a long time, no-one took the time or trouble to remove it.

Then, one morning, it had received a response. 'Yes,' ran the reply, 'But I'm only visiting . . .'

A few days days later, the graffiti was removed . . .

8. THE FORMATIVE MIND: MAN, MIND AND THE UNIVERSE

At the end of Chapter 1, there are three sections, 'Mind: The Soul's Space Suit', 'The Egg of Brahm' and 'The Golden Womb'. It was necessary to place these sections at that point for we have been discussing the role of the Mind, throughout. These three sections, however, are very much a part of our discussion of the Formative Mind and you may at this point wish to refresh briefly your memory as to their content.

THE THREE WORLDS OF THE MIND

It seems to me that every particle and pattern within our physical domain bears witness to organization and design and hence to a designer. Yet there are many who feel that everything has come into existence by spontaneous manifestation and subsequent self-organization. And there are many, too, who never think about such things at all, whose mind are so caught up with the mundane affairs of daily life that such questions of life and death are ignored. This is the nature of our human mind.

But this design was not created long ago and left for us to cogitate and muse upon, but is one *dynamically in the process of creation and manifestation, right now.* We are ourselves a part not only of this creation but also of the creative process and have within us, like all else, the Creator. That is the essence of ourselves and all manifested things.

In the higher regions of the Mind, the astral and causal domains, this eternal presence and creative essence of God is apparent to all souls residing there. Just as we can see that the clouds drift because the wind blows, so is it clear to them that the inward and the outward are linked into one continuum of creative manifestation. They live in a world of being, not a divided world of observer and observed.

In this world, our attention has become so engrossed with our senses and activities that not only have we lost sight of the perception that we are souls, drops of the greater Being, but we have also become unconscious of the manner by which the mind and senses interact. We do not realize that the senses bind the mind into a totally outward perception of material substance and events here. We even forget or become unaware that our human mind is not made of physical substance.

But here and there, the poets, the mystics and the intuitively spiritual humans will experience something of the presence of the Supreme. In fact, many folk possess a background awareness of Something Big, of a power greater than their humanity, but they do not realize the extent to which this experience can be developed. It is for this reason that many inspiring descriptions of such experiences are quoted in *Natural Creation: The Mystic Harmony*.

There is, for instance, a most beautiful passage to be found in Raynor Johnson's *Watcher on the Hills*, though experiences such as this are only a beginning:

> It was one of those gloriously lovely days that one sometimes sees in early summer in England – a cloudless, vividly blue sky with brilliant sunshine. It was morning. The air was shimmering with the moisture from the evaporating dew. I was walking on the lawn looking at the masses of flowers in the herbaceous border. As a gardener I was interested in what was coming up into flower; as an artist I was enjoying the combination of colour, light and shade.
>
> Suddenly, as I paused in contemplation, I was 'lifted' into another world (plane or dimension?). I did not seem to be inside myself though I was still looking normally at the flower border. Everything had become a thousand times more brilliant. Everything also had become transparent. But what was so amazing was the fact that I was not only seeing the colours – I was hearing the colours! Every colour was an indescribably exquisite musical sound, the whole making a harmony that no instruments could produce.
>
> I do not know how long this illumination lasted, perhaps not more than a second or two, but as I came back to Earth, so to speak, I knew that I had been in Reality. The memory of it has remained vivid and real ever since, and has brought me the greatest happiness and understanding. Now I know. I am truly thankful for his gift of knowledge.

In the higher regions of the Mind, the astral and the causal realms, the relationship of the 'inward' and the 'outward' is very clear to the inhabitants. Some folk, who have had no experience of these planes of consciousness, equate the astral and causal regions with states of mind familiar to them as intuitive and spiritually-minded humans, but this is an incorrect appreciation of our true human constitution and of these inner worlds. There are far greater heights than a simple spiritual awareness, to which the soul can rise. A spiritual awareness is only a starting point.

'The entire physical creation,' says Sri Yukteswar, 'Hangs like a little solid basket under the huge luminous balloon of the astral sphere.'

And yet, this is only an attempt to give us some graphic image, for the astral world lies *within* as the immediate blueprint to the physical world. Our world is a crystallized and denser reflection of the inner mansions, or spirit worlds. The fact that we do not perceive them or have trouble imagining just *where* they might be does not alter their greater reality one whit! Just like a child, we need to grow to understand these things.

But many are so entranced and excited by the realization that there is a higher knowledge that they like to feel that they have already arrived. Fortunately, that is not so, and the glories in store for the truly seeking soul are beyond our wildest imaginings and are fulfilling in a way that makes all modern, psychological talk of fulfilment seem like an unwholesome mirage, even a barrier, in the way of such higher mystical experience.

These worlds – causal, astral and physical – are all Mind worlds. They are the three Worlds of the Universal Mind. The causal realm is the world of the finest essence of Mind, duality, time, space and mind energy. The Primal One is barely hidden by the first divisions of the Mind. But division there is and the inner connectedness of all things within the One, is perceived as *causality*. This is the law of the Mind – one of justice, action and appropriate reaction, the law of karma, where pattern gives rise to pattern, and where nothing is really lost or created – only rearranged.

Passing through a threshold, a 'valve' or a 'gateway' – the causal realm is reflected as the astral region whose powerhouse is the *Sahasra*, the *Thousand Petalled Lotus*. These thousand 'petals' or energy currents are formed from above by the interweaving of the three *gunas* (modes of manifestation) and the five tattwas or states (vibrational conditions) of substance, all under the overall patterning or dividing power of the Universal Mind and its prime weaver of illusion – *Maya*. Yet all of these are themselves projected out of the Supreme through the agency of the primal, creative outpouring the Logos, the Creative Word.

Really, all of the increasingly complex patterns of the causal, astral and physical domains are formed under the dominion of the Universal Mind, the prime architect of division, the creator of form.

The thousand petalled lotus, each 'petal' representing an energy current formed as an ex-pression of the interweaving of the more inward energies, is the central organizational plexus at the heart of the astral plane itself and thus of all else below. It is also described as playing the part of the brain in the astral body. Once again we can see how the outward only exists because of the inward, and it tells us, too, how our own physical brain and body are formed, as outward projections of more inward energy currents and how our senses and the world of their perceptions are more closely intertwined with each other and with our mind than we may be aware. But we return to this topic in the final chapter of *Natural Creation: The Mystic Harmony*.

So from the causal is reflected the astral, and from the astral is projected the physical, and in each realm the same essential laws and principles apply, only reflected to a lower condition. In each realm, the soul is surrounded by a mind and a body of increasingly dense material. This is a part of the process by which the creation is projected: the microcosm and the macrocosm are always present. Without the souls and their coverings of Mind, there can be no creation.

It is through the myriad microcosms, each at heart a conscious drop of the Divine Ocean, the creation is brought into being. The Great Being proliferates Himself, but there is never anything other than being. There is never anything other than Him. The principle of the microcosm is the way He does it. From our point of view we can say that we have all jointly made our own creation. From the Divine viewpoint, mystics say that the creation is nothing but the Lord worshipping Himself: the Lord worships Himself through us. But whichever way one looks at it, if the soul-life, the Divine essence, is removed, then the show is brought to an end.

Our human, physical mind, wherein we have our thoughts, however deeply and intuitively inspired, and which gives rise to our physical emotions, is lower than our astral and causal minds, just as our physical body is a reflection of our astral and causal bodies. The higher gives rise to the lower. But the process is not one of linear correspondence. It is far more integrated and interwoven than our linear and divisive intellect can comprehend.

We cannot say, for example, that since we can intellectually appreciate something of the nature of the causal realm, that we have actually reached there, any more than reading a recipe will give us the experience of eating a hearty meal. I mention this because some folk think that if they are able to appreciate intuitively the oneness of everything, they are then very close to actual God-realization. In fact, they may still be a long way even from a limited penetration into the subastral zones. To come to know of the Himalayas does not mean that one has reached the summit of Mount Everest!

Many yogis have spent long and extended lives conquering just the subtle physical aspects of the *pranas* and *chakras*, permitting them complete control of the physical plane, the ability to perform miracles, and so on. But yet they may not have begun their ascent into the true astral zones, let alone the causal zone and the pure spiritual regions above that.

Entry into the astral region gives one full control of the physical realm, though one is enjoined not to use such powers, but to live within the Divine Will. If one does not possess the understanding and consciousness to perform such feats, it can be confidently said that such a soul is still, like the rest of us, trapped in a human body!

When one reads descriptions of the astral region one is struck immediately by the closeness of the inner and the outer. Souls 'travel by way of meditation', they communicate telepathically, their inward thoughts and desires can become immediately and 'outwardly' manifest. A group of them can get together to create conjointly some object and its manifestation reflects the differing ideas which the individuals had concerning it. The statue of an animal, thus created, may be partially like stone and partially like flesh, with its parts comprised of varying and odd proportions according to the mental intentions of the individuals involved in its creation.

More detailed descriptions of this are to be found in *The Secret of the Creative Vacuum*, but consider here these quotations from Yogananda's, *Autobiography of a Yogi*. In this excerpt, his physically departed guru, Sri Yukteswar, returns in his astral form, from beyond the grave, to bring inspiration to his beloved disciple. Bracketed explanatory phrases have been added in a few places.

The intuition of astral beings pierces through the veil and observes human activities on earth, but man cannot view the astral world unless his sixth sense is somewhat developed. Thousands of earth-dwellers have momentarily glimpsed an astral being or an astral world. . . .

Communication among the astral inhabitants is held entirely by astral telepathy and television; quite absent are the confusion and misunderstanding of the written and spoken world which earth-dwellers must endure. Just as persons on the cinema screen appear to move and act through a series of light pictures, and do not actually breathe, so the astral beings walk and work as intelligently guided and co-ordinated images of light, without the necessity of drawing power from oxygen. Man depends upon solids, liquids, gases, and energy for sustenance; astral beings sustain themselves principally by cosmic light. . . .

Luminous raylike vegetables abound in the astral soils. The astral beings consume vegetables, and drink a nectar flowing from glorious fountains of light and from astral brooks and rivers. Just as invisible images of persons on the earth can be dug out of the ether and made visible by a television apparatus, later being dismissed again into space, so the God-created, unseen astral blueprints of vegetables and plants floating in the ether are precipitated on an astral planet by the will of its inhabitants. In the same way, from the wildest fancy of these beings, whole gardens of fragrant flowers are materialized, returning later to the etheric invisibility. Although dwellers on the heavenly planets like *Hiranyaloka* are almost freed from any necessity of eating, still higher is the unconditioned existence of almost completely liberated souls in the causal world, who eat nothing save the manna of bliss. . . .

The fleshly body is made of the fixed, objectivised dreams of the Creator. The dualities are ever-present on earth: disease and health, pain and pleasure, loss and gain. Human beings find limitation and resistance in three-dimensional matter. When man's desire to live is severely shaken by disease or other causes, death arrives; the fleshly overcoat of the soul is temporarily shed. The soul, however, remains encased in the astral and causal bodies. The adhesive force by which all three bodies are held together is desire. The power of unfulfilled desires is the root of all man's slavery.

Physical desires spring from egotism and sense pleasure. The compulsion or temptation of sensory experience is more powerful than the desire-force connected with astral attachments or causal perceptions.

Astral desires are centred in enjoyment in terms of vibration. Astral beings hear the ethereal music of the spheres, [the inward, creative vibration of the Aum, the Logos, the Creative Word], and are entranced by the sight of all creation as exhaustless expressions of changing light. The astral beings also smell, taste, and touch light. Astral desires are thus connected with an astral being's power to precipitate all objects and experiences as forms of light or as condensed thoughts or dreams.

Causal desires are fulfilled by perception only. The nearly free beings who are encased only in the causal body see the whole universe as realizations of the dream-ideas of God; they can materialize anything and everything in sheer thought. Causal beings therefore consider the enjoyment of physical sensations or astral delights as gross and suffocating to the soul's fine sensibilities. Causal beings work out their desires by materializing them instantly. Those who find themselves covered only by the delicate veil of the causal body can bring universes into manifestation, even as the Creator. Because all creation is made of cosmic dream-texture, the soul thinly clothed in the causal has vast realizations of power.

A soul, being invisible by nature, can be distinguished only by the presence of its body or bodies. The mere presence of a body signifies that its existence is made possible by unfulfilled desires.

So long as the soul of man is encased in one, two or three body containers, sealed tightly with the corks of ignorance and desires, he cannot merge with the sea of Spirit. When the gross physical receptacle is destroyed by the hammer of death, the other two coverings – astral and causal – still remain to prevent the soul from consciously joining the Omnipresent Life. When desirelessness is attained through wisdom, its power disintegrates the two remaining vessels. The tiny human emerges, free at last; it is one with the Measureless Amplitude. . . .

The causal world is indescribably subtle. In order to understand it, one would have to possess such tremendous powers of concentration that he could close his eyes and visualize [bring into consciousness, rather than imagine], the astral cosmos and the physical cosmos in all their vastness – the luminous balloon with the solid basket – as existing in ideas only. If by this superhuman concentration one succeeded in converting or resolving the two cosmoses with all their complexities into sheer ideas, he would then reach the causal world and stand on the borderline of fusion between mind and matter. There one perceives all created things – solids, liquid, gases, electricity, energy, all beings, gods, men, animals, plants, bacteria – as forms of consciousness, just as a man can close his eyes and realize that he exists, even though his body is invisible to his physical eyes and is present only as an idea.

Whatever a human being can do in fancy, a causal being can do in reality. The most colossal imaginative human intelligence is able, in mind only, to range from one extreme of thought to the other, to skip mentally from planet to planet, or tumble endlessly down a pit of eternity, or soar rocketlike into the galaxied canopy, or scintillate like a searchlight over milky ways and the starry spaces. But beings in the causal world have a much greater freedom, and can effortlessly project their thoughts into instant objectivity, without any material or astral obstruction or karmic limitation.

Causal beings realize that the physical cosmos is not primarily constructed of electrons, [i.e. subatomic particles], nor is the astral cosmos basically composed of lifetrons – both in reality are created from the minutest particles of God-thought, chopped and divided by maya, the law of relativity [relationship] which apparently intervenes to separate creation from its Creator.

Souls in the causal world recognize one another as individualized points of joyous Spirit; their thought-things are the only objects which surround them. Causal beings see the difference between their bodies and thoughts to be merely ideas. As a man, closing his eyes, can visualize a dazzling white light or a faint blue haze, so causal beings by thought alone are able to see, hear, feel, taste, and touch; they create anything, or dissolve it, by the power of cosmic Mind.

Both death and rebirth in the causal world are in thought. Causal-bodied beings feast only on the ambrosia of eternally new knowledge. They drink from the springs of peace, roam on the trackless soil of perceptions, swim in the ocean-endlessness of bliss. Lo! see their bright thought bodies zoom past trillions of Spirit-created planets, fresh bubbles of universes, wisdom-stars, spectral dreams of golden nebulae on the skyey bosom of Infinity!

Many beings remain for thousands of years in the causal cosmos. By deeper ecstasies the freed soul then withdraws itself from the little causal body and puts on the vastness of the causal cosmos. All the separate eddies of ideas, particularized waves of power, love, will, joy, peace, intuition, calmness, self-control, and concentration melt into the inexhaustible Sea of bliss. No longer does the soul have to experience its joy as an individualized wave of consciousness, but is merged into the One Cosmic Ocean, with all its waves – eternal laughter, thrills, throbs, the longed-for variety in Unity.

When a soul is out of the cocoon of the three bodies it escapes forever from the law of relativity and becomes the ineffable Ever-Existent. Behold the butterfly of Omnipresence, its wings etched with stars and moons and suns! The soul expanded into Spirit remains alone in the region of lightless light, darkless dark, thoughtless thought, intoxicated with its ecstasy of joy in God's dream of cosmic creation.

'A free soul!' I [Yogananda] ejaculated in awe.

When a soul finally gets out of the three jars of bodily delusions . . . it becomes one with the Infinite, [though there are still regions of pure spirit to be traversed before reaching the utterly Eternal].

When Bishop Berkeley maintained that the world was an illusion, Samuel Johnson is reported to have stubbed his toe into a nearby brick. 'It seems quite real to me,' he said. The fact is that from their two points of perception and experience both Berkeley and Johnson were accurately describing how they *experienced* things. These days, with physicists telling us that 'solid' matter is spun into existence out of an energy-rich vacuum, manifesting as an interrelated web of subatomic energy focuses or particles, we cannot be sure about the ultimate reality of 'good old solid reality'. And if we were able to ascend into these higher regions of the Mind, then we would have a vastly different viewpoint concerning the nature of life and of this world. Without that experience or intimations thereof, we are liable to try our best to discredit or discount such descriptions. But like Matthew Arnold's *Hound of Heaven*, the reality will not go away.

Sooner or later we will have to confront it. But the confrontation is individual, within our own selves, like Paul upon the road to Damascus, though it is not usually so dramatic! But it does slowly or rapidly creep up on us. Then it is an intensely personal love affair we develop with the Divine.

THE PHYSICAL WORLD AND THE FORMATIVE MIND

One sees from such descriptions as Sri Yukteswar's – for which, if one cares to research it, one can find numerous independent corroborations – that the astral and causal realms are worlds of higher and more subtle aspects of the Mind. But it is not always so easy to see how the physical plane is also a world of the Mind.

In our physical world things do not seem to be appearing out of nowhere due purely to our mind's activity, unless some miracle is being performed. But how do physical things come about?

All creatures – man, beast, bug and plant – possess a mind. This we have already shown. That mind – instinctive or cognitive, self-aware or unconscious of self – is the source of all activity, of all rearrangements among the energy patterns we think we perceive as this physical plane of consciousness. How does this come about?

Firstly, the physical matter making up the substance of this physical domain is either actually caught up into the bodies of living creatures or has been modified by them in the course of its existence. Even deserts usually become so due to the activity of creatures who once dwelt there, the cycle of desertification often involving man. 'Substance', therefore, at least within the Earth's biosphere has been drawn into countless bodies of living creatures over the course of time. This means that inert 'substance' has been constantly caught up and modified under the influence of mind and life.

Secondly, as we have previously pointed out, all sensory perceptions, whether of immediate objects or of the furthest star are actually subjective – a part of our mind. The physical world actually consists of our mental perception of it. It has no reality to us independent of our mind, even though (of course) things still exist when no one is directly observing them!

Thirdly, when you look at yourself, your relatives and friends, the people you see and meet during the course of a day, and all the other creatures you come across, too, all – without exception – are following the whims and dictates of their minds. Every creature, human or otherwise, has an 'idea' in mind, a desire of one sort or another (even to do 'nothing'!) which is constantly motivating them. Our mind is always busy, telling us what to do next and how to do it. The mind is always active within, in order for any action to be taken or any sense impression to be received.

All of these mind-generated activities and sensory impressions leave their marks upon the soft putty of the mind energy or substance. The motivation underlying our actions as well as the strength of mental, emotional and motor responses induced by our sensory input vary the degree to which our mind energy is impressed. A murderer cannot forget his deed, for example. His revolving thoughts – it is said – even leading him back to the scene of the crime. Our unfulfilled desires and thoughts also leave their impression. In fact, the mental energy field retains a complete memory of all our thoughts, emotions and deeds, in seed form, some of which we are able to draw upon as what we call our memory, while others remain hidden from conscious access, though still deeply influencing our activities.

This energy patterning is retained in potential form, like in the flight-recorder of an aeroplane and – most importantly – it goes with the soul when it departs the physical body. From this black box, along with the same from all other souls with whom we have had association, are fashioned the experiences or destiny of our future lives.

Thus, our mind set, our personality, our physical body, its health and disease, the events and relationships that consume our days and nights – are all spun out from the Mind, in its myriad aspects – individual or universal. We can call this greater Mind, the *Formative Mind*.

So in just the same way that on the astral and higher levels, our 'outer' as well as inner reality is formed out of our own mind energy, so too does the same principle function on the physical level. The magic, however, is *delayed*. What happens *now* has its real cause in the mental impressions gathered *in previous lives*, and those, too, were linked to the past. So the mind goes on and on, and all souls revolve, willy-nilly, on the wheel of birth and death, under the influence of their individual minds.

It is this formative level of mind function which is ignored in most modern psychology. But it is only from such understanding that a framework or paradigm is provided whereby the brain physiologist, the neuro-scientist, the psychiatrist, the psychologist and the physicist can all talk coherently to each other. At the present time, they have practically nothing useful to say to each other. Clearly our modern scientific descriptions of how a human being is put together are grossly lacking in some essential way.

The brain, as I described in *The Web of Life*, is the elaborator of mind function at the physical level. The brain, indeed the whole body, needs to be approached as an image projected upon a screen from an inner mind pattern. This is an energy model where the physics of manifestation through the vacuum gateway also needs to be understood, too. Psychiatry and psychology can therefore be shifted onto a new ground where activities are seen as energetic patterns within energy fields, rather than as mental concepts, fabrications of the intellect. In fact, we need to develop a science of *psychophysics*, the energetics and physics of mind function.

THE BRAIN-MIND INTERFACE

We 'live, move and have our being' in a cosmos comprised entirely of being, of experience. Even our knowledge of what we call the brain is only a sensory experience – which itself lies in our own mind.

What we call our body and brain are thus only sensory experiences or perceptions within the multi-faceted realms of our own being. There is nothing external to us. There is no objective, outward substance. There is only being. This understanding leads to a fascinating insight into the nature of the brain-mind or body-mind 'interface'.

For it means that *there is no interface as such*! All we have is being. That being is comprised of soul 'surrounded' by mind. Soul, here, being synonymous with 'consciousness' or 'life force'. *So the brain is only the sensory experience of our own human mind.* What we call the brain is only our way of experiencing mind at the level of sense experience. And sense perception is an aspect of mind and being. In fact, for the most part, that sense perception lies in the visual field. We *see* the brain, rather than *hear* it, for example!

This by no means invalidates scientific study of the brain, but it does put such study in a far wider context and explains why one can never find thoughts and emotions within the brain. For the brain, however detailed the analysis, is only an analysis of *sensory* experience. Thoughts and emotions lie in another area of experience altogether! Thought experience and emotional experience can never be found nor understood through a study of sensory experience!

Of course, there will be parallels between brain and mind, for we are one integrated *being*. Hence the neuro-endocrine connections and other relationships discussed in *The Web of Life*. Also, one tends to talk of things as if there were an interface in a vertical or inward energy spectrum. But actually, that 'vertical spectrum' is nowhere outside us, but is only the multi-faceted realm of our own being.

THREE LEVELS OF HUMAN MIND FUNCTION

Since the days of Freud, our perception of human psychology has acknowledged that our mental-emotional experience is a mixture of what is conscious and what is subconscious. What we think of as our conscious mind is really only the surface of our subconscious mind.

Our mind does not possess its full potential for awareness because it has moved out (e-motion or ex-motion) from its centre behind the eyes – the thinking centre, or eye centre. It has moved out because the karmas of our destiny, etched into the fabric of that human mind itself, have impelled it to do so. Thus we act unconsciously, without understanding of how and why things happen in the way they do, or how and why we think and feel like we do.

This substrate of mental causality, giving rise to both the mental-emotional as well as so-called outer experience, is included in what I have called the Formative Mind. From the point of view of human psychology, there are thus three levels of mind function, each based upon the more inward existence of the other. To restate it, these are:

1.   'Conscious' Mind
2.   Subconscious Mind
3.   Formative Mind

We have pointed out that in our normal human lives, it is not possible for us to know how or why things happen as they do. Sometimes, the coincidence of events is such that we intuitively know that there has to be a hidden connection. We call it synchronicity, serendipity, coincidence – even chance or luck. But from the higher point of view there is no such thing as chance. All is planned, owing to the prior content of our mind. This is how it is that the same patterns repeat for each of us, time and again, pleasant or unpleasant. It is why some folk are always lucky and others continually unlucky. The mental, karmic trend and pattern is already set, already coloured, and we live through that. Just as our basic personality is identifiably retained throughout our life, changing in only very rare instances such that a person becomes unrecognizable, just so do our destiny karmas, the so-called outer events of our lives, have a colour or trend which mirrors that of our inner mind. The outer events happen only because of the inner mental content; they are part and parcel of that.

Psychologists recognize this, but they do not always understand the energetic basis and the multilevel function of the mind, which makes all things happen, although both Jung and Freud spoke of psychic and mental energy. Not everyone 'believes' in reincarnation, and this severely hampers understanding of mind energy function, for the process of a human life is then envisaged to commence with a mystery, for which there is no cogent explanation.

Child prodigies who can play concert piano or tournament tennis aged four or five will always remain anomalies in any concepts which exclude the mental influence and tendencies of previous lives. Indeed, since personality, temperament and intellectual abilities are clearly not transmitted genetically, the mystery of why all of us are born with personalities and talents that are a part of us throughout our life also remains a mystery. The origins of our personal psychology and mind set have no explanation unless one countenances the possibility of previous births and the continuation of mental tendencies from one life to another, or the fulfilment of the desires of one life in a future incarnation.

These factors, then, are all a built-in part of the process of mind manifestation at the physical level and to help us understand these processes, the delineation of the Formative Mind is, I think, a useful one.

THE FORMATIVE MIND AND PHYSICAL LIFE

It is the Formative Mind, for example, which is the real power behind positive thinking. For it is that which produces both the positive outlook *and* the 'outward' responses and events to mirror it. Folk appreciate this formative aspect almost unknowingly when they realize that one's *attitude* is of great importance in shaping (or *forming*) all the affairs of life.

Similarly, clear mental *vision* has been a companion to all great achievers in this world. Often, the vision or goal held firmly in the mind is of an obsessive nature. One could not suggest that Alexander, Napoleon or Hitler after being born 'normal', decided early on to do what they were going to do. From their earliest days, they were inwardly constituted in that way. Life worked out that way for them and their powerful mental visions became reality, for some limited period of time.

Similarly with great artists or musicians, their artistic expression would have been a compulsion with them, all their life. Many were child prodigies. Their desires from past lives all focused on that one life and becoming a master of their craft was the result. But the fruits of their work is never from beyond their own mind, however inspired they may be. The personality of the composer, the mind of the artist, these are impressed into every musical pattern and every brush stroke. A scene from Monet or Constable, or a symphony by Beethoven, Delius or Mozart is almost immediately recognizable as the work of those particular artists or composers, since they present us with an identifiable reflection of the individual creator's mind.

Indeed, it can be said that our lives and circumstances have in effect, *already happened* at the time of our birth. We only have to follow through the promptings of our inner mind for the apparently outward events to come to us, for the two are closely linked into one. Yet within it all there is enough, highly conditioned 'free-will' for us to create new impressions upon our minds: new karmas, to set up the scenario for future lives. If we have any free-will, it is – most importantly – to struggle against negative tendencies, to change our inner mental attitude, to develop a more wholesome and spiritual outlook – and to make the effort not to create new karma.

TEAMWORK, DEMOCRACY AND THE FORMATIVE MIND

In this world we all appear to have individual, separate minds. Therefore, whenever, within a group, something needs to be organized – whether 'important' or 'trivial' – there is apt to be confusion. Each individual formulates his own plan of action according to his own personal circumstances and desires – and the result is frequently chaotic! As the old joke goes, 'the definition of a camel is a horse designed by committee'. 'Free' social structures, therefore, have instituted the democratic process of government, or social organization by a consensus of opinion. This is the nearest that intrinsically divergent and separate minds can get to unity of purpose!

Within specifically defined frameworks, however, individual minds can be trained to work in synchrony. This we call teamwork, and it can be found in everything from sports teams, to industrial and commercial effort, to music, to social groups and in family life. But the key to such teamwork usually lies not so much in consciously attuning to or actually developing genuine awareness of the minds of others, but in developing joint mental habits, either by active training as in sports teams and the military forces, or by social habituation, as in families and communities.

Situations and activities then develop along habituated and largely unconscious lines. As a means of creating integration, this works, but the extent to which such a team can function is limited by the conditioning or training to which the members have been exposed. There is often little room for spontaneous creativity.

The fact is, however, that at the deeper, formative levels all our minds are already linked. At the deepest Mind level, they are united in the Universal Mind. In the astral and causal regions, therefore, the souls – still encased in Mind – operate while knowing themselves, increasingly, to be 'parts' of one great whole. The democratic process, therefore, is not required for things automatically happen in an integrated fashion. All minds automatically work together to manifest the most appropriate outcome, in harmony with the Divine Will flowing through them and all things. This is co-operation in its highest expression – something rarely found in this world, dominated by the egocentric human mind, its resulting self-centred activity, and its recalcitrance in realizing its own intrinsic oneness with all others.

MIRACLES, CLAIRVOYANCE, TELEPATHY, SUBTLE PERCEPTION AND THE FORMATIVE MIND

The nature of the Formative Mind explains too, how miracles are performed. For when we realize that what we think we are perceiving outwardly is only like an image on a screen, projected from within our own mind, then to alter the 'outward' image on the screen, we need only concentrate within and alter the patterning in the deeper levels of Mind energy within us, from which the image is projected. So miracles require only concentration of mind, the making of an inward and upward shift in the level of our attention.

Just as all souls are one in the Divine Ocean, so, too are all minds one at the level of the Universal Mind. This mental oneness is also reflected into our human minds. Our karmic associations with each other from past lives, etched into our individual minds provide the seeds of our destiny in this life such that we all become co-creators or shareholders of our mind-dream in this physical world.

*pos. thinking

So miracles occur from a consciousness of the inner levels of Mind function. Similarly, are telepathy, clairvoyance and so on, a part of this deeper level of mind function. Clairvoyance, or knowledge of the future, comes when the patterning of destiny, already etched into the hidden energy matrix of the human mind comes into the realm of our conscious attention, or conscious mind.

Mostly, our minds lack real inward concentration and so the concentration unwittingly comes and goes. Therefore, clairvoyant perceptions of this mind level often come as *flashes* of insight, or inward perception – as indeed do intuitive and intellectual insights – rather than as a continual ability to read the inner mind patterns. Additionally, because of human weakness – imbalances within our mind energies – these perceptions are usually clouded by our own emotions, desires, personality and sense of ego. Consequently, they vary in their veracity.

The clearer, the more humble, and the more concentrated the mind, the more readily and more accurately will its images be readable, within its own self, just as – conversely – the reflections in a pool of water are disturbed when the surface is ruffled or agitated by passing wind and weather.

Similarly, telepathy automatically arises in those who possess a more inward and spiritual nature. In essence, telepathy is the awareness that all our minds are inwardly linked, together with the subtlety of mind making it possible to 'read' this linkage to some degree. As one might expect, telepathic attunement is commonly found between close members of a family or between close friends. The mother, for example, may know intuitively when a child is in difficulty, even though they may be separated by greater or lesser distances. So telepathy is actually no more than an awareness of the integration of all our minds at the deeper, formative level – an integration which is present whether we are conscious of it or not.

These associations with each other are all a part of the karmic outworking from within the Formative Mind and so we are more readily attuned to those people with whom we are attached or associated, for the intertwining of our destinies is written into the mind centres of all concerned.

We may find, too, that our mind is ahead of our destiny, so to speak, and the thought of some friend pops into our head out of 'nowhere', only minutes before they appear around the corner of the street or the telephone rings. Or we may know – because their mind is reaching out to ours – just who it is who is telephoning, but before we pick up the handset. This is all the functioning of the Formative Mind.

All events in the world are registered at this level and an understanding of Formative Mind function explains many phenomena and experiences previously considered separate and inexplicable.

Many of the tribes who live close to nature have never lost this knowledge from past, more spiritual ages, and they use such communication automatically. Those left at home know when the hunters have found game, or when someone is hurt or in difficulty. When unsure of their direction they may send out their 'dream wanderer', travelling on their etheric or perhaps astral form to find out where their fellow humans are. They may know, too, of events happening far away in the world, even those of little practical concern to themselves.

In addition to a superb range of senses, animals also perceive the subtle side of things, though they do not possess the cognitive powers which humans can bring to bear. Yet it is these very cognitive and intellectual powers, plus the tendency of our mind to go out, away from its inner centre, which makes us either forget we ever had such telepathic and subtle faculties or makes us doubt their existence even when we have the experience. We need to use and understand our intuition, learning also to discriminate intuition from imagination.

We only have to think of our close relationships to know how we get locked into each others' mental atmosphere through the intermingling of our thought currents. Then we often think the same things at the same time, even if physically distant – in different rooms or in different parts of the world.

Though there is no doubt that we are connected when physically separated, all the same there is most definitely a powerful mental energy which has effects at close quarters. It is a part of the aura we each carry with us, reflecting outwardly the content of our inward minds, as subtle energetic emanations.

Thus, some folk bring a sense of peace, while others carry with them a surging, gusting, vibrating energy of uncontrolled and unconscious thought and emotion. This is part of what influences crowd behaviour, either to the negative extremes of anger and even physical violence, or the positive, calming vibrations of beautiful music or the nearness of a true holy man, whose presence sheds almost tangible waves of love – spiritually uplifting those lucky enough to be inwardly touched by his vibration.

The actual content and structure of this greater or Formative Mind can be *directly explored and experienced* in the higher forms of meditation, though it must be said that a competent and experienced guide and teacher is essential before setting out to explore the untrammelled inner spaces of one's own being. Then, of course, the means by which miracles can be manifested becomes quite clear, though one is instructed never to use such powers, since they increase the ego and waste spiritual power.

The unfoldment of one's own destiny, including the time and manner of one's own death, then becomes like an open book to the initiate. Indeed, the experience of death is undergone in full consciousness and understanding, even with a sense of utter relief that the physical dream-life is over.

Such experiences can come upon one unexpectedly and may radically change a person's outlook upon life. These include spontaneous mystic experiences, lucid moments and near-death experiences, the origins and nature of which I discussed more fully in *The Web of Life*.

In some out-of-the-body and near-death experiences, the individual sees their whole life pass before them, even events which had been long forgotten. It has been said throughout history, too, that the drowning man sees his whole life float before his eyes. Though I have always supposed that this could only have been known from those who *nearly* died – not those who never actually returned!

This experience once again demonstrates the nature of the mind as an energy field in which all its past actions are stored, as well as its future destiny. The past forms the future in an endless circle of holistic causality, like the circular track of a bulldozer or a tank, which proceeds in an apparently forward direction by endlessly going round in a closed loop.

Folk who have relived their old memories in this fashion and then been restored to physical life have found that those memories were a higher experience in their own mind, one far more vivid than any normal daily recall of past events would have been. Their attention is withdrawn totally into their own mind and there, in its formative substratum, they come face to face with the mental record of their life's journey, as faithfully and unemotionally recorded as in any scientific instrument. This is our own personal akashic record, and it stretches back over many lifetimes, too.

PRAYER, DESIRE AND THE FORMATIVE MIND

The power of prayer to manifest things or events also comes about through the functioning of the hidden energies of Mind. All our minds are connected at the formative level. Thus, when we *pray* or simply *desire* for things to be a certain way, we set in motion unconscious waves or currents in the Formative Mind which sooner or later makes things happen that way, according to the intensity of our prayer or desire. But it is the *mental* intensity and sincerity or focus which makes things happen. Mere repetition of set words while the mind wanders elsewhere has little effect.

Such prayers or desires, however, inculcate attachment to the outcome and to the object of the prayer, so the net result is to keep us tied down to the level of our desires or prayers. Many people pray for the health and welfare of others. This seems commendable, but we are most probably using our own limited spiritual power to set such healing energy in motion and while another may gain, we become the loser.

Prayer for enlightenment, for grace and for inner mystic understanding or for guidance in our affairs of the world would seem to be alright, for these humble the human mind and prepare us for inner ascent. Sometimes such prayer inculcates the response of our own higher mind and sometimes we may even be guided by beings from more inward spheres who 'hear' or perceive our prayers as Mind vibrations, and can help us, according to their own spiritual stature. But really, the highest 'prayer' is meditation with a focused mind, for this represents the soul's aspiration to reach God and speaks louder than the mental repetition of any verbal prayer, standing beyond desire for material things.

But mostly, in prayer, we only contact our own higher mind, which may even project to us a vision of some personage, usually long since dead, in whom we imagine we have faith. Thus, many visions of past saints or saviours are the projection of our own higher mind. Such visions may be blissful and even tell us of events which come to pass, thereby gaining our confidence. But all this lies within the functioning of our own inward layers of Formative Mind. They arise therein, just as the characters in our dreams are entirely fabricated out of the impressions lying within our own mind, disappearing in the light of the higher waking consciousness.

MIND OVER MATTER

Every great sportsman knows that games are won as much in the mind as in the body. This is true of all achievement. The great heavyweight boxer, Mohammed Ali, knew it when he attempted to bolster his own self-image as well as intimidate his opponent long before the contest actually began. The body is only the outworking of the Formative Mind, in more ways than the most obvious. We do not have conscious awareness of how we even bend a finger, let alone perform highly co-ordinated tasks. The area of mind-body linkage is usually lost in a cloud of unconsciousness. The Mind, with all its subtle aspects, provides the blueprint or prototype for all bodily and physical forms. In this sense, the mind is not a part of the body; rather *the body is an aspect of the Mind.*

Every sportsman – indeed every one of us – has good days. Days when our mind is sharp and clear, our energy flows more freely and with confidence. We forget ourselves, even; yet we are more concentrated than usual. Actually, the little 'I' or ego is an illusion of our thoughts, our human mind. Confidence and free flow of energy from within-out come when we forget that little 'I' in a deeper state of mental concentration. The self-conscious person is awkward and ill-at-ease, as we have all experienced. He also makes wrong decisions out of an incorrect appreciation of the situation.

Some games are almost entirely of the mind, like chess. The tussle is fought mentally with only token moves of a physical nature. It is the mental meaning and intention behind the moves which is of relevance. Similarly, the mind-to-mind and generally mental nature of modern communications and information technology all demonstrate our movement towards a more mental way of being.

Living and being in this world is more an attitude of mind than an amalgam of the acts we perform, for the acts follow naturally from our condition of mind, and our happiness or otherwise lies entirely in our mind. These days, sportsmen even seek specialist help of a psychological nature to help them win. At least one British football team has hired a psychologist who specializes in helping players overcome particular weaknesses in physical style.

His approach begins *off* the field. Having identified a particular weakness, say kicking to the left with the left foot, he makes the player rehearse the move numerous times, but *in his mind*. And the player rehearses it with great finesse and panache, as if he were the world's greatest expert at it, paying particular attention to the specific manner in which he does it. This forms a deep pattern in the player's mind. And since it is the mental patterns at an unconscious level in the Formative Mind energy which enables us to perform any outward action, the mental repetition of an image in which the player can really perform the action well is automatically replayed when the player meets the situation on the field. So his ability is improved.

Practice makes perfect; but while the muscles and body structure need to be supple and well exercised, without the correct mind patterns lying behind them, nothing is achieved. Practice creates a habit in the mind, which can also be engendered even in the absence of physical practice.

This is the role, too, of *desire* or *vision*, the mental pattern within, which makes things happen outwardly. As the song says, 'A dream is still a dream, if it's still inside your mind.' But ultimately it does come true, though we may need to take another birth for its outward manifestation to be arranged.

Back with the sportsmen, the book, *The Psychic Side of Sport*, by Michael Murphy and Rhea White, is a fascinating study based upon the authors' conversations with innumerable top athletes and players. And many of them concur and have experienced the way certain players actually *will* events to happen about them, in the way they do. There are even cases of well-known personalities, out of the strength of their desire and mental concentration, spontaneously *willing* the ball (or whatever) into performing unusual gyrations – against the 'laws' of physical nature.

As we have seen, this is actually how yogis, ascetics and others can consciously perform miracles after practising their meditation for long periods of time. Meditation is nothing but a natural way to concentrate the mind within, on its own inner structure, thereby uncovering – sometimes unwittingly – the way in which the physical universe is manifested out of the Formative Mind. The ability to perform 'miracles' is thus acquired.

This physical world is actually a Mind world, and we collect mental impressions unintentionally, as well as consciously, from every side. We pick up the vibrations, or thought currents, directly from those with whom we associate or meet, even casually. We are influenced by the general atmosphere of places where we live, work or visit – the keynote, so to say, of the Formative Mind patterns which have created and are creating that place, and are drawing that kind of person to it. When we 'feel at home' in a place, we stay. If we are out of tune – our mind patterns do not resonate with such a place – we want to leave. Like gravitates to like, in the world of the mind.

'When a liar meets a liar,' says Maharaj Sawan Singh Ji, 'They like each other. But when a liar meets an honest man, their relationship snaps.'

Then, too, we carry our individual impressions with us, from past lives. Our inclinations, our talents, our interests are all derived from the mental impressions first seeded in past lives. Now they sprout and flourish for their allotted span.

The same applies to other creatures, too. This is a hidden factor not usually appreciated in laboratory experiments performed to determine animal skills. Successive generations of rats learn to run mazes faster, for example. Even rats in a distant laboratory perform better when others of their kind, quite unknown to them, have solved the problem first.

There is, most probably, as Rupert Sheldrake suggests, a spreading of the skill as a morphic pattern in the keynote of 'rat'. The ratty current within the Formative Mind 'spreads', reaching out to all rat-kind. The morphic pattern is actually a mind energy pattern. This happens with man, too, as we all ride the same karmic Mind wave, so why not with rats, at their particular resonant 'frequency' or pattern?

But it is also possible – though it has to be after earlier trained rats have died – that those same old rats had come back as their own great, great grandchildren, carrying the submerged impressions of previous maze-solving skills with them.

Somewhere, in the back of their ratty mind, they have done it all before and it only needs a little trigger to rekindle the flame of the ability, for the dormant seed to sprout. This is also how young children of our times pick up computer techniques so rapidly. They are a part of today's formative, karmic mental wave and may have worked with such things in past lives, too.

Indeed, it is likely that the staff of one laboratory will unconsciously generate close association with one group of rats, with the rats taking birth as rats time and again within just that one laboratory. So naturally, that group of rats will learn tasks with increasing rapidity because they have done it all before!

A group of scientists in another part of the world may therefore get different results when re-performing some experiment because their rats will not have the training of past lives built in to their mind set!

Rats and all creatures, have a destiny, just like ourselves, spun out from the inner recesses of their own mind patterns. Both the experimenter and his rats have a karmic assignation which they cannot avoid! And no two meetings will be alike, for all minds are different, even among rats, not to mention experimenters.

Researchers tacitly acknowledge this. This is why they use more than one rat in their experiments, to get some sort of average, and why scientists like to repeat other scientist's experiments. Genetic differences in the rats are only secondary to the mind's structure and patterning. If we could understand rat psychology, we would find that they were all different, just as any dog or horse breeder will tell you that his trainees have different natures or personalities, as we humans do. Certainly, even twins and litter mates follow different destinies, this itself expressing the difference of the more inward patterns of the Formative Mind.

It is all a part of the individual mind patterns with which a creature is *born*. Yet no geneticist has ever found one scrap of proof that psychology and mind patterns are encoded biochemically into the DNA. No scientist has ever found the slightest scrap of real evidence that any true mind function, in beast or man, is encoded *directly or only* into the brain or central nervous system. The reverse is in fact true. The brain is *involved*, no doubt, but it is not the prime mover.

Yet, since the conventional scientific paradigm refuses to acknowledge the existence of anything other than material substance, the belief persists that mind is somehow to be understood as patternings of physical matter. In spite of all the evidence to the contrary. The implications of the alternative perception are more than most materially minded people can tolerate.

BODY AND MIND

We have seen that all our sensory perceptions of this physical world are ultimately found only as subjective experiences in our minds. And since the world does not disappear when we cease looking at it (or otherwise perceive it with our five senses), it also means that there is a stratum of Mind at a deeply unconscious yet collective level which holds the physical universe together. More correctly, the physical universe is a part of this Formative Mind.

We can take the matter still further, however. We are only aware of our physical bodies through these same five modes of sensory perception. Are our physical bodies also, then, only subjective experiences within our mind? It certainly seems that way.

So one can say that even from the point of view of sensory perception, our physical bodies are a part of our mind. Not only is a physical body the most outward layer in the expression or crystallization of the hidden hierarchy of Formative Mind energies, but even our minute-by-minute experience of our physical body, being sensory, also lies within the more conscious part of our minds.

So it seems that in whatever direction we turn, our mind is involved – either at a level at which we are presently conscious, or at the deeper, formative level of which we normally possess no consciousness. There is no 'substance', no objective world as such, only being. No wonder scientists have a problem understanding the nature of the brain-mind interface! For all our sensory examinations and observations of the brain and mind, as well as intellectual and intuitive thoughts concerning them, lie within our mind itself. Mind is trying to understand brain and mind. Where does one start? The answer to that one can only be: direct experience of one's own inner being and structure. And that means: meditation.

THE FORMATIVE MIND AND BODILY PROCESSES

The bodies of all life forms are projected images of energy patterns from within. They are images on the screen of space, the energy of vacuum, the substratum of all physical existence. The pattern maker within is the Formative Mind, energized by the soul, consciousness or life force.

Looking at the outward appearance of human and other bodies, we see certain recurrent shapes. We observe head, thorax and abdomen as common traits throughout all vertebrates and something akin to it in most invertebrates, too. There is always a central control system equivalent to the head and spine, a respiratory and heart area, and an abdominal area concerned with the assimilation of food. Even unicellular creatures and protozoans have nuclei which act in the same central organizing manner as the brains of higher creatures.

Again, creatures have organs of motion and manipulation – arms, legs, wings, fins, scales, cilia and their similars. All creatures higher than the vegetable kingdom possess some body structure enabling them to get about of their own accord.

Among vertebrates, we have varying numbers of toes and fingers, though normally never more than five on any one limb. We have protective nails on fingers and toes. We find horns, hooves and claws. All performing more or less the same basic functions. There are the wonderful feathers of birds and the scales of fish and snakes.

Internally, we find bones, hearts, intestines, livers, spleens, gall-bladders, ovaries and testes, male and female sex organs. The same kind of organs are present throughout all creatures, from insects and worms, to whales, primates and man.

Going deeper still, we find nerves, blood vessels and an endocrine system present throughout. We find the same cellular structure even in plants. Indeed, in a plant, some cells are elongated, stretched between its leaves and its roots like long pipes, designed to carry nutrients and fluids between its upper and lower extremities, an indication of a polarization of function and therefore of subtle energy structure. All atoms, all cells, indeed all of nature is bi-polar, as I have discussed in previous books. The cellular structure of bodies does itself bear witness to the polarization inherent in all Mind regions which gives rise to discrete and organized structures under the influence of threshold and pattern-forming processes.

Within each cell, we find a nucleus containing DNA, the primary centre of organization and administration. We find the mitochondria, prime ministers to the nuclei's presidency.

Reaching further into the configurations found in these amazing mosaics we call bodies, we discover a complex integrated patterning of energies which we, in our intellectual minds, divide up into the dynamic patterns of molecules, atoms and subatomic particles. We divide the dancing, vibrant tapestry into ions and electrons, into the electromagnetic radiations 'emitted' and absorbed by every molecule, into the magnetic fields associated with electron spin and hence integral to the functioning of every atom in the body; into the oscillations and vibrations of all these integrated aggregates of energy; into the four primary forces of modern physics. All these are dynamic, shifting shapes, patterns and rhythms we may perceive in this projected image of the body at the level of analysis defined by the physicist.

Taking the matter deeper still, we can understand that all these forces, molecules, atoms and subatomic particles are only spinning, whirling dances in the vacuum energy ocean. They are the bubbles on this amazing sea we understand so little. They are only observable patterns in space.

Then we realize that *all the shapes and patterns which we have previously been examining are only effects, only reflections of patternings or blueprints deeper within the structure*, which have at last eluded our physical grasp.

Moving inwards, we find ourselves in the subtle physical realm of chakras, pranas and mind substance. Here lies the blueprint from which the outer creature as we perceive it with our senses, is formed. It is from here that behaviour and instinct are determined, providing – automatically – as a reflection of what lies within, the sense organs, the limbs, the DNA, the cell and organ structure, the molecular and bioelectrical patternings – everything with which we are presently familiar and much more which we have yet to uncover.

But this subtle, physical-mental blueprint is itself not fundamental.

It is only a reflection of what lies deeper within. Thus can the hierarchy of energies be traced higher and higher, through astral and causal levels and beyond, until we reach the inward Source of all, even beyond the Universal Mind, the Ocean of Life and Consciousness. This is the true creative essence within all life forms – the Ocean of Life itself.

The spark of life within all creatures is the soul. All creatures in their most inward essence are souls, sparks of the Creative Life Force. For this reason alone, we should respect and revere the life within all forms. When we deprive other creatures of their physical life, we mar the balance and subtle qualities of our own life. It automatically affects our own mind, growing an ever-hardening crust of lower consciousness.

When we take their dead bodies into our own for its maintenance, we automatically link our minds to theirs, to their suffering. In the intricate, hidden and interlinking processes of the Formative Mind we become responsible. The energy field of our mind is tainted and marked, our 'sin' is entered in the great book of the Mind, as with all our actions and thoughts. This is part of the karmic process, for which we will later be called to account: our mind energy will automatically find expression in this and future lives.

So all bodies are the end result of a multifaceted inward process of blueprinting or creation. *Creation is happening right now*, not long ago in some cold creative act of the Divine, after which He went on holiday. The Divine Puppeteer is present in every tiniest part of this creation. He is aloof and yet universally prescient. Even the physical processes of continuous manifestation out of the vacuum state, the perpetual motion of subatomic particles, bears physical witness to this.

The intricacies of life forms, spun out of inert matter, are only maintained while a spark of the Life Force, modulated by mind energy structures, is present within them. When the soul or life departs, death ensues and the body immediately begins to disintegrate. The continuous presence of the life force and individual mind within is required for the continuous integration of bodily processes.

Perceiving just the outward similarities between species, evolutionists have assumed that the higher have evolved out of the lower. How else, they say, could such similarities be present within all life forms? But one might just as easily suppose that one atomic element has 'evolved' out of another because of their essential similarities in construction. Or that salt has evolved out of sugar. Similar appearance does not necessarily imply evolution from the one to the other.

It implies a similar process of construction or design. It suggests similarities in the underlying blueprinting mechanisms. It does not suggest an evolutionary process.

But conventional evolutionary theory is a mechanistic and totally physical approach to the understanding of life. If one begs the question of where all this world of material substance has come from, or how it is being continuously maintained in existence, then one has eliminated all higher or more inward possibilities.

But one cannot leave such questions unanswered. In fact, the similarities between all creatures exist because we are all children of the same Father, all sparks of the same Divine essence The inward processes of this creation are similar for us all. He has created the Mind and all its multifarious energies. According to our karmas, according to the patternings within the Formative Mind, we receive a physical body spun out of the five tattwas and patterned by these inward essences of energy. This process is described more fully in *The Web of Life*.

The level of consciousness or awareness which this subtle and gross physical vehicle permits a creature, depends upon its completeness as a reflection of all the spheres of the creation. Man alone possesses the ultimate microcosm, the full reflection of the entire creation, within and without. Only man, therefore, has the potential to realize God, though He is still present within even the lowliest bacterium.

The energetic patterns of mind and matter surrounding the souls of other creatures are only partial microcosms. This is what makes them what they are. Thus they may often be in awe of man, even frightened. They do not have the capacity for our degree of awareness.

Yet, all the same, because the inward essence of mind and subtle energies are of the same nature, so we find similarity in form at the physical level. This is the point at which we began this section. From the shape of heads and limbs, from wings and fins, from hearts and livers, from the similar function of blood, nerve and muscle cells, from repeated molecular patternings and processes whether of bacteria, birds, beasts or men – from all of these we can deduce that there must be a common ground of inward patterning.

It is only differences in the inward blueprints which constitute the essential difference between all creatures. It is the patterns of the Formative Mind which determine how the outward projection of the creature actually appears to our physical senses.

HEALING AND THE FORMATIVE MIND

The understanding of the body as a projected image of energy patterns from within leads one to a radically new appreciation of the nature of healing – conventional or alternative.

This paradigm of what a physical body really is allows us to understand many things which were previously obscure, many of which I have discussed in this and my previous books.

We have to stand back and see things from a perspective in which the body is both a designed, continuously maintained manifestation, *and is also, therefore, only an effect*. An effect of Formative Mind activity fuelled by the inward soul. This entire physical domain and all the bodies in it, is really only a world of *effects*. The *cause* lies deeper within.

So physical healing deals only with physical effects and, in some cases, subtle blueprints. This is the origin of psychosomatic conditions, though actually everything concerning a body is psychosomatic, even bending a finger.

Any kind of healing or medicine must realize that it starts from a point of ignorance. We may, according to certain definitions of our intellect, map out some part of the processes by which this projected image of the body appears to function. This we call medical science. But we are really only performing a process of image analysis not an exploration of the inward design and manifestation process. We are only looking at how the pieces of a jigsaw puzzle fit together, without ever knowing how the jigsaw was conceived and made.

Healing often means trying to re-pattern the image without understanding how the image comes to be upon the screen. The screen, in this case, is the vacuum state, the energy of space, the physical akash. If we do not understand how what we call subatomic particles arise upon this screen, then all our subsequent analysis of the way these particles behave as atoms, molecules, bioelectricity, cells, organs and whole bodies is entirely suspect. It is based upon foundations which we have taken so much for granted that we have failed to ask the fundamental questions: '*What* is it that we are actually examining?' and, '*Who* is it that is performing the examination?'

Hence, especially when the motivations become those of financial and egocentric reward, we study our bodies in a most remarkably deficient and naive fashion. We would never approach the research of molecular substances – drugs – in the way we do, if we understood the depths of our ignorance. Much drug research is entirely empirical – 'Suck it and see' – a clear indication that we do not understand the basic underlying, formative principles at work in our bodies.

This is not a matter of criticism, for we humans are born in ignorance. The tragedy is that we are unaware of the depth of that ignorance. We do not realize how much a human being can really come to know through inward study and direct perception. We even ridicule those who, having glimpsed beyond the curtain, try to tell us what they have seen.

However, being pragmatic, and accepting our state of ignorance, the only approach remaining to us is to work with nature. To use the faculty of our higher mind, our intuition (but not too much imagination!) to guide us. To open our minds to subtle influences and intimations, to be quiet and humble, to use our human discrimination, to quieten our mind and intellect.

As humans, we may never understand the real how and why of nature's forms and patterns – whether molecular, bodily or in the world at large – but we should attempt to come into tune, into resonance with them. This has been the hallmark of all the great natural philosophers and physicians. Then we can find many ways of helping ourselves and our fellow creatures.

BODY LANGUAGE AND THE FORMATIVE MIND

Mind, motive, behaviour and body are all linked. This is the origin of body language. Mind expresses itself automatically and mostly unconsciously through the body and its posturing. This is a necessary part of the natural economy, permitting creatures to read each other's intent. Were a wolf to be really dressed in a lamb's body and demeanour, then he would be the prime hunter. He could walk up to any potential prey exhibiting the innocence of a young herbivore, before – at the last moment – revealing hostile gnashers with which to decapitate his unfortunate prey.

The balance of nature would soon be disturbed if predators were always successful hunters. By Darwinian logic, such duplicity should perhaps have become a law of nature, for it would give all predators the edge. But no – the mind always and automatically expresses its intent. The slinky, hunting cat, or the purring contented pussy by the fire – these two distinct feline mental moods are eloquently expressed in the cat's demeanour. It cannot operate otherwise. Mind always precedes and accompanies action, automatically exhibiting itself to the observant eye – human or otherwise.

The worlds of the Mind are intended to be places of interaction and participation. The creation is meant to be a place of ex-pression, where what is within is revealed automatically in what is without. This is true both of our physical mind and body, as well as in the creative process of the higher worlds. The outer is always a reflection of the inner. To perceive this mechanism in action is a question of consciousness. Even a man whose motives are other than his conduct can be rumbled by one who has clear perception. Many a dog has proved an excellent judge of character, better perhaps than his 'owner', barking furiously at one who 'looks fair, but feels foul', in an attempt to alert all and sundry to the scheming and negative intent of the one they suspect.

PHYSICAL BODIES AND MENTAL MICROCOSMS

We have said that a physical body is a projected image, part of an inward microcosm, comprised essentially of subtle Mind energies. For ultimately, the tattwas are themselves of Mind origin, forming in their first seed form in the causal realm of the Universal Mind.

But this projected image differs from an image on a screen, in that it is also a dynamic, two-way, energetic structure. Not only is the outward a reflection of what lies within, but changes to the outward form also affect the inward structure. It is a two-way process, a shimmering dynamic projection of Formative Mind energies underlain by the power of the Life Force.

So the life or soul within all creatures is the same. The differences lie in the microcosmic mind structure which surrounds the inward Life Force. The more this microcosm contains reflections or points of reference to the macrocosm – both the inward and the outward universes – then the higher is the potential consciousness, awareness or intelligence of the creature.

And this applies not only to creatures of this physical world, but to those of the subtle physical, the astral and the causal realms as well. A 'creature' of course, means one who is created.

The human form is the only form containing the full complement of microcosmic centres. Only man has the potential to contain a consciousness of the entire creation within himself. Only man can consciously realize God.

The 'lower' creatures, possessing less than complete microcosmic mind and body garments, are limited in their perceptions and ability to grasp the nature of existence, of the Whole.

Why are some souls born into human bodies and others into the various forms of the lower creatures? This again happens due to the law of the Mind, of karma. The mind 'earns the right' to a particular mind-body structure, a particular microcosmic configuration, according to its own past activities, its 'worthiness'. The law is automatic and inexorable. We go where our mind – our karmas – take us. The myriad forms of life – both present and in the fossil record – are only expressions of the various forms which the subtle and shifting mind can exhibit. As the great sea of Mind shifts its emphasis, so too do the outward forms. This is the secret to understanding 'evolution', discussed at length in *Natural Creation . . . Or Natural Selection?* Nature is perfectly exact. The mental tendencies and the available physical forms match each other perfectly, at all times. There is a tailor-made, physical 'box' for every shape of mind – for the Mind actually forms the bodies in which it dwells. The dove-tailing of mind and body is more than exact – it is intrinsic in the way creation manifests.

This is why saints have called this world a labyrinth, a maze, *chaurasi*, as it is known to Indian mystics – the wheel of eighty-four – that is of 8,400,000 types of 'species'. This is why they suggest that having been born at long last as a human being, the only point whence escape is possible from this physical whirlpool, we take advantage of it and try to effect our escape. But this, they say, is possible only with the help of a mystic who has himself made good his escape, or – in some instances – may never have suffered the apparent indignities of chaurasi, but have been sent direct from God on a mission of mercy.

We only receive what our mind desires. To have received a human form is a rare occurrence. This becomes clear when one thinks of how many other souls there are, on this planet alone, incarcerated in other bodies. Since the distinguishing characteristic of the human form is that the trapped soul has the potential to become God, it also means that we do not receive such a gift without there being some desire in us, somewhere, to understand that greatest mystery of all, the secret of life itself.

We must, therefore, do our utmost to keep this spark, this inner quest, fresh in our mind, to entertain the highest spiritual aspirations. Only then can we hope to continue forward in our spiritual progress. Otherwise, we may descend once again into lower forms.

As long as we continually aspire to the highest we can comprehend, then the onus is upon the Lord to direct our footsteps towards Himself. Maybe it will take many births before the inward desire for Him is polished and brightened up, so that we may ultimately be ready to meet a true mystic. There may be many stepping stones on the way in the form of religious, spiritual, moral, ethical, or social philosophies and activities before our minds are pure enough to appreciate the simple and universal majesty of a perfect master. 'When the disciple is ready, the guru appears,' says the old aphorism. But we are also led to that point of readiness. We are inwardly moulded and guided so that ultimately we do recognize a master when we meet him.

Prior to that, he may be living just close by, but we feel no attraction to him, nor understanding of his teaching. Our mind is still too heavily engrossed in the outward display of activity. But we can only do the best we can, according to our own understanding. The rest is up to God. And the Great Spirit, the great inner source of Life is interested in every spark that takes its origin from Him. He is not unmindful of His creation. He knows His purpose, though it may be obscure to us.

MULTILEVEL EXPLANATIONS AND DESCRIPTIONS

One of the interesting aspects of the Formative Mind is that explanations of the same phenomenon or event can be given in many different ways and from many different levels. Once one understands the creative process, then the level of any required analysis is largely determined by practical considerations. It depends upon the level of perception from which one decides to view things.

One should note that this is quite different from the 'levels' one considers when talking of the 'molecular level' of enquiry or the 'atomic level' or the 'subatomic level'. In this case, the 'levels' actually refer only to an increasingly detailed analysis or observation at the sensory level of experience. So all these 'levels' are at the *same level* – the bottom or outermost! The levels of Formative Mind energy are increasingly subtle blueprints of each other and provide the means for the manifestation of the more outward levels. The more outward are created out of the more inward. They are levels along an *ontological dimension*, a dimension of being within our own selves.

So depending upon the purpose one has in mind, one makes an analysis from whichever of these levels is the most appropriate. For example, when dealing with violent criminals, one has to be quite mundane and practical. Such socially destructive behaviour needs, quite simply, to be curtailed, and we lock the person in prison. But the reality is that we are probably dealing with a mind set which perceives such behaviour as perfectly acceptable or at least as an irresistible compulsion. We could also argue that we are all partially responsible for their behaviour, because of the social conditioning and pressures which we have all jointly created. There can be no thieves if there is nobody to steal from.

So while safeguarding society from such a person's *physical* activities, it would make good sense to move up a level and try to help the person to a more harmonious state of *mind*. In the process, the idea of punishment, moral judgement, revenge and so on, evaporate as we attempt to see things from a mental level. By moving our point of perception, we thus develop a more compassionate outlook.

Finally, from a *spiritual* standpoint one can say that the unconscious suffering and inner unrest felt within the disturbed mind of the criminal individual will ultimately – perhaps after many lifetimes – drive the soul towards God and a search for the great inward source of being within, where all mental conflict ceases.

So all three ways are quite valid in their respective spheres and quite compatible with each other in the context of practical living. There is no paradox if the full context is understood.

Similarly, we do not need to understand subatomic physics and the fundamental forces of nature in order to make bricks and build a house. However, it does most definitely benefit us, even in the mundane affairs of daily life, to appreciate the full context of all our activities, for then we work in harmony with the natural 'laws' of the creation. And the full context means an understanding of the spiritual and Formative Mind dimension to everything.

But perception of this 'context' is a matter of consciousness, awareness, intuition and experience. It is not something arrived at by intellectual analysis alone.

Indeed, it is often quite impossible to understand things at a purely physical level, because Mind factors are involved in answering the questions we are asking. This includes all of medicine and healing, for example, for the organization of a body comes into being because of the Mind and the inner life force or consciousness. Similarly, in a study of the fossil record, we must introduce the Mind factors if we are to gain anything like a clear perception of what has been going on. We cannot fully understand anything unless we realize the role of this greater Mind within the projection process of creation.

INTERLUDE

VIEWPOINT ON CREATION

The following is an extract from a letter written by the great mystic, Maharaj Sawan Singh Ji, to an American disciple, during the early years of this century.

There are two ways of looking at this creation:

1. From the top, looking down – the Creator's point of view.
2. From the bottom, looking up – man's point of view.

From the top it looks as though the Creator is all in all. He is the only Doer, and the individual seems like a puppet tossed right and left by the wire puller. There seems to be no free will in the individual, and therefore no responsibility on his shoulder. It is His play. There is no why or wherefore. All the saints, when they look from the top, describe the creation as His manifestation. They see Him working everywhere.

Looking from below, or the individual viewpoint, we come across 'variety' as opposed to 'Oneness'. Everybody appears to be working with a will, and is influenced by and is influencing others with whom he comes into contact. The individual thinks he is the doer and thereby becomes responsible for his actions and their consequences. All the actions are recorded in his mind and memory, and cause likes and dislikes which keep him pinned down to the material, astral or mental (causal) spheres, according to his actions in an earlier life in the cycle of transmigration. The individual in these regions cannot help doing actions and, having done them, cannot escape their influences. The individual acts as the doer and therefore bears the consequences of his actions.

As stated above, the observations differ on account of the differences in the angle of vision. Both are right.

1. The individual clothed in coarse material form sees only the external material forms. His sight does not go deeper than that.
2. If he were to rise up to *Sahans dal Kanwal*[1], the same individual would see the Mind actuating all forms. The form would be only secondary; Mind would be the prime mover in all.
3. The same individual, from *Daswan Dwar*[2], will see the Spirit Current working everywhere, and will see how the Mind gets power from the Spirit.
4. From *Sach Khand*[3] the whole creation looks like bubbles forming and disappearing in the Spiritual Ocean.

Maharaj Sawan Singh Ji (Spiritual Gems)

[1] The powerhouse of the astral plane, also called *sahasra*.
[2] The first region of pure spirit, lying immediately above the Universal Mind.
[3] Literally, True Home – the Eternal 'Region', the Kingdom of Heaven.

9. THE FORMATIVE MIND: ORDER AND THE ENERGY DANCE

SYNCHRONICITY, VIBRATION AND PATTERN

Something I have mentioned in my previous books is the phenomenon of apparently separate, yet identical or similar developments taking place simultaneously. This happens frequently in scientific circles and is a well-known, yet unexplained, phenomenon. Similarly, we find social and philosophical attitudes changing in synchrony throughout the world. Carl Jung, who was fascinated by it, called it *synchronicity*.

One could say that the phenomenon arises because people, whether in science or in general attitude, are simply responding to similar stimuli. This is no doubt, to some extent, correct. But there is more going on than this. We are all travelling on one wave. If you watch a wave on the shore, it breaks at many points simultaneously, yet these breaking points are, at their outset, separated from each other. But immediately they begin to spread and merge until ultimately the whole wave is embroiled in the energy of its breaking. It is the processes *within the wave* that cause the breaking. Breaking is a part of the 'being' of the wave at that time.

In the same way, all the waves are interconnected, too, with each other. One forms behind the other as a part of the progression and as a response to the other waves. It is also a cyclic pattern. And beneath it all, lies the great ocean from which all waves and wave crests arise and within which all are connected. Similarly, we have all conjointly created this world – in a far deeper sense than we may realize. We all have a powerful formative part to play in the interwoven tapestry of energy change and patterning that makes up our lives – both individually and in a planetary sense. We are all connected like wavelets on this vast ocean of karmic activity, within the control and administration of the Formative Mind.

So that great bug-bear of modern physics, the incidence of apparently non-local connection, is an integral part of the continuous manner by which the subtle energies of our minds weave the matrix of planetary affairs. The processes of the more subtle energy domains are inherently more interconnected than we observe in the outer world.

This we see quite clearly in the national characteristics of a country or area where the mentality or mental characteristics of the individuals are reflected in the environment. This has an obvious explanation when one considers, say, the architecture and design of man's artefacts, for they are naturally a reflection of the minds of their creators. And those minds are conditioned by the past and respond to the present to produce a changing yet coherent pattern of change in the design of buildings and so on. Conversely, the nature of a place affects the minds of its inhabitants in both obvious and subtle ways. Those living on the coast, for example, are inclined to an interest in fishing and boating, while those living in the mountains develop a rugged mentality to suit, as well as a love of their grandeur.

But this line of horizontally causal or linear thinking does not seem to give us the complete picture when we consider that actually the whole of nature responds as one. There is, so to say, an integrated keynote or vibration that influences an area. The environment, the creatures dwelling there and the general thought processes of the people are all integrally intertwined in one pattern.

In China, for example, even the trees have that wonderfully intricate, twirling structure that is reflected in the Chinese mentality – the fine detail, care and subtlety which one finds in their art, in their social customs and in the graceful design of their buildings and gardens.

The Chinese will even nurture a stone. It is tended, lying in the river bed, by generations of family or community members, until at last it is brought in to take up its place in the garden. And with all that care and loving attention woven into its energetic substructure over a period even of centuries, the quality of its vibration is greater by far than if it had been bought at a supermarket garden centre and carried home for immediate installation on the patio! The underlying mental motivations and attitudes in the two situations are so entirely different.

Similarly did the Chinese tend the shape of their hills and the design and siting of all their artefacts. And the result was one of great beauty and harmony. I have discussed this Chinese science of

Feng Shui in *Subtle Energy* and this understanding of the way in which atmosphere is built up is – or used to be – an intrinsic part of Chinese culture. Though the influence of modern times is now taking its toll.

Again, the exquisite Ming vases from the fourteenth to seventeenth centuries are still considered of great beauty and finesse, throughout the world. And both the silkworm and the ability to weave and dye its delicate silk into strong fabrics were of Chinese origin.

So this intricacy and detail in the mind patterns of the Chinese and many of their creatures are reflected in all their environment. Both man-created and otherwise. Because it is the minds within ourselves and all other creatures that fashion our environment, both in the more obvious as well as in more subtle ways.

This was brought home to me very clearly during a trip to the Swiss Oberland one spring, some years ago. The beauty of the mountains peaked in snow and covered in wild flowers, enclosing valleys where the farmers allowed the hay-fields to be filled with a profusion of wild blooms was entrancing. All the common species found in England were allowed to grow and no doubt add nutritional variety to the hay crops. Ragged robin, forget-me-nots, wild parsley, wild carrot, red campion, cranesbill and many more made each field an inspiration that Monet could never have passed by. And the human surroundings showed evidence of caring hands, manifested even in the charming architecture of the farm buildings. There were very few excrescenses and disharmonious human attacks upon the all-encompassing vibration.

Returning home, fired with enthusiasm to recreate my English lawn in the image of a Swiss hay-field, I soon realized, in less than a few minutes' survey of my own small plot, that what had been so entrancing in Switzerland was the subtle atmosphere and vibration that *gave rise* to that multitude of flowers. The power of the mountains, those great towering masses of rock and crystal, the sheltered valleys, the way of life – all these and much more of the indefinable had gone into the creation of nature as it manifested in the valleys of Lauterbrunne, Wilderswil and Wengen.

England is England – a different vibration with different subtle and mental characteristics. So I saved and collected wild flower seeds and have tended them with care. I have planted them into my 'lawn' and created a piece of the softness and gentle joy of an old English meadow. This is the harmony and supportive vibration of England. Nonetheless beautiful, but different, all the same.

So these integrated waves, wavelets and breaking crests are an inherent part of the way nature manifests. I can readily imagine, too, how – given certain common conditions prevailing throughout our planet – the same or similar species could have come into being apparently independently in different parts of the globe, bearing characteristics that reflected both the local vibration as well as the planetary circumstances and subtle atmosphere. So, palaeontologists could be wrong in assuming a geographical connection when they find fossils of similar-looking creatures in different parts of the globe. They may not have migrated. They could have arisen apparently independently, but connected by the creative patterning of the Formative Mind which interpenetrates all living forms.

For the real creature lies in its subtle mental structure which reflects outwardly in its physical form and behaviour, just as we do ourselves. And it is the changes in this subtle atmosphere to which species must surely be responding when 'evolutionary' steps are made. Without such a change in vibration or atmosphere, then there is little change in form or behaviour. When there are greater changes in the quality or characteristics of the formative mental vibrations, then greater changes will occur in outward form, too.

Surely species do not change by chance, by random mutation? They flow with the changing patterns of environment and vibrational atmosphere. There is nothing random in nature. 'Not a leaf moves, but by His order,' said the great Indian mystic, Nanak. 'The hairs of your head are all numbered,' said Christ. The great, Universal Power is within everything and is responsible – as one whole – for all change and all patterns in creation.

HOLISM AND THE FORMATIVE MIND

Holism, as a philosophical outlook, arises automatically in the mind from intuitive depths which we may not comprehend, whenever an individual's degree of consciousness begins to expand. Then we are not so gripped by the linear machinations of our habit-ridden intellect and begin to see that there is a wholeness in nature, and a wholeness in ourselves, too.

Wholeness in diversity is a characteristic of the Formative Mind. The Egg of Brahm, the Golden Womb, is one whole dynamic energy system of Mind. The physical creation, what we call nature, is just a small part of the whole. The oneness, or the wholeness, comes from God. That is the primal, uncreated Power or Source of energy, of consciousness. The mind may make patterns, but the oneness is never lost.

In this physical world, we see only effects, so we perceive the inherent wholeness and oneness as an intuitive awareness of holism as a principle in nature. We see limited causality. No prime causes of this physical spectacle are open to view. Inasmuch as our soul and higher, more inward aspects of mind, are functioning, we see a greater and greater degree of Oneness. The less our consciousness, then the more divided is our perception. Then habit, intolerance, egotism, close-mindedness, materialism and a blind adherence to religious, social, philosophical and scientific dogma take hold of our mind and intellect.

Such a mind acts subconsciously to defend itself against all ideas which smack of the universal, the mystical, the spiritual, the holistic. It reacts, without ever knowing how or why it does so – or even that it *has* done so. Ignorance often masquerades as authoritarianism or cynicism. Few are the scientists and philosophers who admit to lack of fundamental knowledge. The 'expert' or 'authority' may be only an unconsciously adopted facade to hide the ignorance inherent in being human. 'Stiff in opinion, always in the wrong,' as the old adage has it! The free-thinker never executes the intolerant. Yet give the intolerant a little power – and then watch the heads roll!

ORDER AND THE LAWS OF NATURE

The mixture of philosophies stitched into modern science is quite astounding. We have discovered the highly ordered structure of DNA, the amazing complexity and integration of processes active within just one cell, the fabulous workings of bodily biochemistry and physiology where everything happens under extreme *organization* and by *design*. But then, say some of the scientists, this patterning came into being and responds to changing environmental circumstances through *random* mutations.

It is thought to be random only because the real causative pattern is invisible to us. We do not even understand the nature and origins of causality. But because many scientists, like the rest of us, do not like to admit to their essential ignorance, the most bizarre theories are put forward to fill the gaps in our understanding.

The situation is identical in the probabalistic aspects of quantum theory, as I have discussed in *The Secret of the Creative Vacuum*. Man discovers order in all things, he seeks out the *laws* of nature, but where our perceptions fail us, we immediately involve chance and random fluctuations as fundamental explanations. We are indeed an amazingly inconsistent species!

Indeed, we neither understand the fundamental basis underlying either the deterministic, mathematically defined laws of nature, such as gravity, nor those describable by 'laws of probability' or statistics, such as quantum theory. In fact, since both represent our observation of patterns, relationships and causality, the real origin of which is unknown to us, there is really no conflict between the two. Mathematics, geometry and statistics do no more than describe patterns. But the manner by which the patterns come into being is beyond our normal ken.

If man's science were even half a science, it would have predictive capabilities way beyond its present limitations. Man may be able to predict and determine the outcome of a few linear pathways – the falling of an apple, the sending of an electronic module to the moon – but when it comes to integrating a knowledge of all the parts into understanding the functioning of the whole, our logic is proved impotent. We cannot even predict what will happen to us in the next two minutes!

Some physicists may think that they are on the verge of solving the riddle of the universe and yet they have totally neglected the nature of their own mind, life and consciousness. They do not know the nature of the 'I' who is attempting to gain that universal understanding. Biologists may think that by analysing some of the linear pathways within living organisms they understand the nature of life. But yet no-one knows how to begin the creation even of the 'simplest', single-celled organisms. And the entire area of instinct, behaviour, and how it *feels* to be some other species is totally neglected. Conventional biological science has no idea *how* such instinct and behaviour comes into being. Man does not comprehend his own mental structure, let alone that of the other species.

Yet, forgetting that he has so little knowledge of his own being and mind, man thinks that this unknown and uncharted entity, the human mind, can tell him of the secrets of the universe. He is placing total reliance upon a faculty, the real nature of which is quite obscure to him. He is using an instrument to tell him of the ultimate verities, yet he has never even calibrated his instrument or come to know what it really is.

It is beyond the ability of outward science and philosophy to fully understand these things. Nature cannot be really understood by analysis. The infinite whole cannot be known by the finite, logical analysis of the fraction of a per cent of this whole that actually comes before our mind and our sensory perceptions. And our mind and consciousness, too, are a part of this whole. How can the part know the whole unless it becomes the whole?

These things can only be really known through inner exploration within the laboratory of the human body. Through mind and consciousness, man the microcosm is able to access all parts of the universe and to understand them, by finding them within his own being, his own consciousness. This is the only way that the part can know the whole. Our reductionist science of analysing the parts has a place within the scheme of things. But it is not its part to ever understand the whole by such thinking. The mind and soul have higher faculties for obtaining such knowledge.

MATHEMATICS, GEOMETRY AND THE FORMATIVE MIND

Mathematics and geometry are simply a language for the specific description of relationships within this great tapestry. They do not represent the reality itself, however, any more than the *word* 'orange' is the orange itself. Moreover they are, ultimately, unable to fully portray just how the One becomes the Many and yet remains within every part of the Many. They cannot fully represent the hierarchical, multifaceted, multilayered structure of the Formative Mind. They have great value, *but* their limitations must be fully comprehended.

Everything in our physical life has its limitations and the intellect as a means of understanding 'life, the universe and everything' is most definitely one of them. So we should learn to use it, but with a clear perception of its natural limitations.

At the present time, mathematics and scientific descriptions are mostly used to represent relationships in a purely horizontal manner – analysing the patterns and rhythms we observe in the screen-image through our senses and call the physical universe. Sacred geometry (a study of relationships in the physical universe as expressions of the One becoming Many) and the mathematics of some vacuum state theorists attempt to add in the ontological dimension of the Formative Mind. But at present, such work is quite rudimentary.

A detailed version of sacred geometry and mathematics, for example, should be able to tell us how the universal 'constants' come to possess the values they do and how the mathematically and statistically defined 'laws of nature' actually arise. This requires analysis of the Formative Mind projection system to tell us how the image upon the screen actually comes to possess the shapes and rhythms which it does. When looking at physical bodies, we should be able to show in exact detail how organisms come to possess a repeating cell structure with identical DNA within each cell nucleus; why the kidney is kidney-shaped; why DNA is DNA-shaped; and indeed how the whole dynamic, rhythmically shifting and integrated pattern of a body comes into and is maintained in being. This would be a *real* model of a body, human or otherwise. A starting point for this kind of multilevel energy model is attempted in greater detail in *The Web of Life*.

So it is imperative that we understand this vertical, formative dimension and the real role played by the Formative Mind in the creative projection process. Without this perception we become lost in the detailed analysis of an image, without ever realizing the role played by the deeper Mind within all of us. Then we miss the crucial point that sensory observation and perception not only *require* our mind, but that the things we think we perceive outside are actually a part of the greater, Formative Mind. That the physical universe is a Mind world where all its laws are 'laws' of the greater Mind manifesting at a physical level. We thus fail to realize that the physical universe is a joint Mind creation in which all soul-mind-body entities are involved as co-creators or shareholders. That is, that *being, life or consciousness* is the essence of all existence whether of 'inert' substance or living creatures.

At the deeper level, we are all drops of one great ocean with the potential to realize our intrinsic Oneness with that Ocean. But the Ocean is of Being and Consciousness itself. There is absolutely nothing *inert* or lifeless about it at all.

So scientists are utterly misguided if they assert that life, being and mind have arisen from material substance. The true situation is exactly the reverse: material substance comes into being as a crystallization of Mind, something associated with each and every soul, human or otherwise.

This greater Mind appears outwardly as individual mind, but at the deeper level, all mind substance is interwoven and includes what we call physical matter or energy.

We do not immediately perceive it as such, because the direction of our attention is downward and outward, away from the eye centre. We thus dwell in unconsciousness or subconsciousness, for we increasingly lose consciousness when we live below this centre. But when, by the correct practice of meditation, concentration is withdrawn back to this eye centre, or third eye, then we begin to perceive directly the way in which the universe is constructed. And this is an experience of being far more satisfying than any intellectual or philosophical description. But quite beyond any conventional scientific verification!

THE LAWS OF NATURE AND THE FORMATIVE MIND

Trying to describe the physical universe in purely empirical and physical terms is, therefore, like describing the plumbing system of a house by assuming that water arises spontaneously and under pressure, at the main water inlet. Or it is like describing a domestic electrical system under the assumption that electricity arises at the main electrical input.

In both cases, as long as the character and properties of the water or electricity are defined exactly as 'universal laws of nature', then the remainder of the domestic plumbing or electrical system can be described with ease and empirical consistency. And the description enables us to do a lot, in a practical way.

What it does not permit us to do is to understand the *intrinsic* nature of water and water pressure, or electricity and the electron. It does not tell us what they are. In fact, in the excitement of unravelling the ramifications of domestic plumbing and electrical wiring, the relevance of this more fundamental question is likely to get forgotten or deferred.

Similarly with our perception of the 'laws of nature'. We can, by empirical study, come to understand much concerning the way in which natural forces *behave*. But this tells us nothing of how such 'laws' *arise*. *How* do 'universal constants' come to possess the values they do? *Why* does $e = mc^2$? *Why* does light move at 186,000 miles per second in a vacuum? *What* are mass and energy? *How* do the mathematical laws describing gravity, optics and electromagnetism arise? In short, how does the physical universe come to exist in the way it does? And *why* should these mathematically definable 'laws' remain constant? In fact, *do* they really remain constant over aeons of time? If we do not know how they arise, how can we be sure that they do not exhibit a periodicity, just like everything else in this physical universe?

Such questions will always remain unanswerable if we perceive the physical universe as the only objective reality, forgetting the fact that we, the observer – our mind and consciousness – are integrally involved in its existence. When we take the image on the screen as the only reality and forget to ask who is the 'I' who perceives this image and what is the role of that 'I' then we can never answer the fundamental questions. When we realize that physical existence is inextricably bound up with the nature of our own being – our own existence – then we begin to move in the right direction.

Then we begin to observe that behind every form there lies the greater, Formative Mind. Just as in human relationships, we increasingly understand that to truly appreciate other people we have to come to an understanding of the way their mind works, so too in the whole of creation. Behind the whole pageant of rhythm and pattern lies the Formative Mind.

In our daily lives, we intuitively know that to perceive and judge other people based entirely upon observation of their outward physical behaviour and appearance, tells us little of the inner mind of the individual. We have to look to the psychology and inner life of the other person to know the how and why of what they do, outwardly.

Similarly with the entire physical creation. If we ignore the nature of the primal energy and the patterning power of the Formative Mind, we can never know how either we ourselves or material substance actually come into existence. If we ignore the projection system, we can never understand the how and why of the image upon the screen – however much we may analyse it. But when we realize that there is a projection system in which we the observer are inextricably bound up, through the Mind, then things begin to fall into place.

THE SCIENCE OF THE SUBTLE: SUBTLE BIOLOGY AND
SUBTLE PHYSICS

We are in the habit of thinking that form, structure and organization only exist in our bodies and in the substance of the 'outside' world. But, as we have seen – and please forgive the repetition – this patterning arises in the energy hierarchy of the Formative Mind. We have therefore to realize that our subtle energy fields, our inward mind energies, also possess form, structure and ordered patterning. A study of this is a science one could call *Subtle Biology*.

But the means of actually perceiving this structure requires a movement of our point of perception up the dimension of being. There are psychics, for example, who can actually watch *from within their own mind*, the functioning of their brain and how mind and brain 'interface'. Or they can see the subtle patterning energies of their own or other people's bodies and thus diagnose disease by direct perception.

Good, clear-minded and honest psychics with reasonable control over their imagination and ego are thus excellent people to have around in laboratories, or in medical and healing practices! Direct perception of this inward formative dimension of being and energy is very useful.

In *The Web of Life* I briefly discussed the state of consciousness of *idiots savants*, generally called autism. Here, we find individuals whose normal mental contact with the physical world is almost non-existent and yet some of them possess the most amazing talents. There is one, for instance, who can hear complex music and immediately reproduce it on the piano and guitar. Another can memorize the intricate lines of ornate and complex buildings, reproducing them later in detailed graphic form. But he draws, not by first drawing the outline and gradually filling in details, but by starting in one corner or at the top and drawing the whole thing as one might unroll a sheet of paper. It is as if the mental image is present before him, within his mind, and he is simply copying it on to paper. A perfect photographic memory requiring great natural powers of inner concentration.

Yet another can tell you, within a few seconds, the day of the week of any date in the past or the future you may care to ask. 'The third of June, 1428' and back comes the answer, 'It was a Tuesday' – or whatever. Yet the man cannot even add up correctly. Clearly he is using mental faculties unfamiliar to most of us.

Just as we recall and use words of our language without knowing how we do it, so do these *idiots savants* have special and unusual talents which they cannot explain. Clearly they are operating from a different point on the dimension of being than the majority of us, using their mind in a manner different to ourselves. I did discuss this interesting and revealing phenomenon in *The Web of Life* and will go only a little further here.

Demonstrating even more clearly my original point that the subtle energy fields possess a formative structure and organization is the *idiot savant*, Daniel, from Canada. Daniel's forte is making electronic toys. But his methodology is bizarre! He simply sticks a transistor here, a resistor there, a capacitor somewhere else, a bulb in one corner and a switch in another. He does not even wire them together. They appear to be randomly glued on to a piece of perspex. Yet when he switches them on, the bulb lights up. In fact, when *anyone* switches them on, even when he is not in the room, and he is involved with something else, they light up! So any constant psychokinetic influence from Daniel's mind is ruled out.

Clearly, Daniel can see, *in his mind*, the inner structure of energy patterns at the subatomic and vacuum state levels. Unhampered by preconceived ideas concerning what is and what is not possible, and working along the mind energy hierarchy into physical manifestation, he is simply rearranging the energy patterns of physical manifestation to do his bidding. From his point of view, he is simply playing with his toys and wants them to light up! And with both a direct mental perception *and* manipulative capability, he arranges the structure of the vacuum state and its manifested subatomic particles to take on the patterns he desires.

Perhaps, although he cannot explain himself as we might like, he is also trying to tell us something. The enigmatic and elf-like smile of the man who can tell you the day of the week for any date certainly seems to indicate that he knows that he is baffling and confounding your everyday conceptions of what is real and possible. But he cannot explain himself.

So in Daniel's toys, do we have the forerunners of twenty-first century technology, of applied *Subtle Physics*? Clearly we have to learn a great deal more concerning the structure and patterning of the vacuum and more subtle energy fields, to be able to do the same thing by non-psychic means. But it does demonstrate a little of what might be possible if we could understand how physical energy comes into manifestation as an aspect of the Formative Mind.

Presently, of course, the allied sciences of subtle biology and subtle physics have yet to be taken seriously by the vast majority of scientists. The times, however, *are* a-changing.

Some of the scientists who have witnessed Daniel's toys in operation and even taken them to their laboratories for testing, have been reduced to tears, stating that all that they had been taught and believed in appeared to have been turned upside down.

This is not strictly true, however, for when we understand that what we call the laws of nature are only *partial descriptions* of reality, not inherent in nature itself, then we can see that there is always room for expansion of these descriptions. Einstein did not prove Newton *wrong*, just *relative* to a particular set of observational circumstances. Similarly will all of our present science, Einstein and all, be proved relative to its position, the lowest rung, on the ladder or dimension of being and energy. The times *are* a-changing!

The implications of Daniel's demonstration are immense. Our present electronic technology relies upon the creation of various effects, ingeniously linking these effects together (we call them circuits) and manifesting yet further effects – heat, light, sound, motion or effects used as logical triggers, switches and memories in computers and electronics.

But we are playing with electrons and transmuting energies without our ever knowing what they actually *are*. No physicist knows what an electron – or electricity – actually *is*. No description or theory, however exact, can ever convey the real nature of the thing itself.

Being autistic, Daniel is quite untainted by our scientific conceptions. So when he looks at a transistor or a light bulb, for example, he sees its characteristics and potentialities as configurations of vacuum or spatial energy. And when he takes a piece of perspex and glues his various components into place, he is connecting up various dynamically active spatial patterns. Where necessary, he arranges, aligns or structures the vacuum and subatomic energy patterns of the perspex and probably the components, too, if they do not quite do what he wants of them. He does this *with his mind*.

What Daniel's gift is telling us is that to create the various effects we humans want – light, sound, locomotion and so on – we could perform with far greater finesse if only we could structure or order the vacuum or spatial energy according to our needs.

This is what we are doing anyway with all our wires, components and miscellaneous bits and pieces. How much neater it would be if we could use these subatomic particles themselves – these spatial energy vortices – as well as the more hidden vacuum structure as our circuit, as our energy pattern transducers.

MORPHOGENESIS AND THE FORMATIVE MIND

All of the phenomena and patterns described by Rupert Sheldrake in his book, *The Presence of the Past*, are explicable in terms of the Formative Mind and an understanding of creation out of the vacuum state. For the Formative Mind is itself the pattern-maker, following its own integrated and naturally holistic laws. Mind is always formative. That is its inherent nature and manner of operation.

From the Universal Mind and the causal realm, to the physical human mind and its outward expression as the physical body, Mind is the inward blueprint. Whether we are waggling a finger or observing the forms and structures in our bodies and the rest of nature, the Formative Mind lies behind the forms and activities we perceive. Whether we are trying to understand the subconscious psychology of ourselves or of our fellow humans, whether we are dealing with psychosomatic effects; even when we are brought face to face with the miraculous and the supernatural – it is always the Formative Mind at work.

If we do not understand its operations, it is because of our own reduced level of consciousness, our own level of perception, and therefore we are confused. Clarity and direct perception arise when our consciousness expands. Then, slowly and slowly the way in which it all happens becomes clear to us. As we discover how we are put together, we automatically discover how the 'universe' is constructed, for the one is a part of the other, our human form being the most amazingly constructed microcosm. Then the true nature of what the 'universe' really is also becomes clear.

SUPERNATURE AND THE FORMATIVE MIND

Similarly, the amazing and fascinating panorama of facts concerning the natural world arranged by Lyall Watson in *Supernature, Supernature II, Lifetide* and his other books, are all explicable when one understands the multilevel, multifaceted and totally integrated aspects of the Formative Mind.

Lyall Watson himself suggests, as does Rupert Sheldrake, that the Collective Unconscious of Carl Jung is a part of this picture. But the Collective Unconscious is usually thought of conceptually, as are many ideas in physics. In the world of the Formative Mind, everything is a matter of experience, energy and being – something you can actually perceive and experience as a formative, creative power in creation. Something you can (figuratively speaking) stick in a jam jar! But the laboratory for this experience and observation is within our own mind and life consciousness. It is no use trying to analyse sensory experiences in order to understand things of the higher Mind.

Supernature is the Formative Mind itself, the hidden patterning spun over and around the inward life principle. At the physical level, we see only fleeting glimpses and surface phenomena of an inwardly dimensioned whole of incomprehensibly vast proportions.

Similarly, the great order, integration and 'coincidence' identified by James Lovelock in his *Gaia Hypothesis* and Dr Brandon Carter in his *Anthropic Principle* are readily understood when one realizes that the physical universe is only an image upon a screen, a dynamic projection or crystallization of more inward Formative Mind patterns. But these two revealing perceptions of order and design in the physical universe are discussed more fully in *Natural Creation: The Mystic Harmony*, for they represent clear scientific evidence in favour of the Formative Mind hypothesis.

THE WEAVER OF THE WEB

So the Mind is always formative. This is its intrinsic nature. To divide is to create patterns. To form patterns is formative, per se. Whether we are formulating plans in our 'conscious' mind for the things we desire or wish to do, or whether we realize the influence of our own subconscious mind upon these 'conscious' thought and emotional patterns – the patterning, Formative Mind is always at work. First we have the mental idea, the blueprint, then the action takes place. Remove this blueprint or thought process, as in unconsciousness, and we fall over in a heap, or slump in our chair.

But our human thoughts, however mysterious they may be are only a part of the totality of Mind, and the same formative process continues throughout all Mind worlds. Our destiny – the events of this life – are a part of our physical mind energies. It is from here that past mental impressions form the destiny of the present life and that patterns of the present are later formed into the destiny of future existences.

The powerhouse of the astral region, the Sahasra, the Thousand Petalled Lotus, is the formative energy crossroads for all lower creation. But it is itself only a pattern of energy, reflected from the more inward, causal realm. This causal region, the world of the Universal Mind itself, is the primal pattern maker, the originator of Maya, the weaver of the web of illusion, the spinner of the beguiling tapestry of forms which we are led – by our minds – into considering so real.

But Mind itself only exists due to the power of consciousness, the Ocean of Being, of God, the Primal Power or Energy. The life-giving Light within this magical Pandora's box of forms is the Divine itself. All Mind patterns whether of Universal Mind, of our human thoughts or of anything within the causal, astral and physical universes, are only bubbles upon this great Creative Ocean. Mind is a creation of the Supreme Consciousness.

What we call energy is thus only patterns in the Creative Life Stream induced by the Formative Mind. And since consciousness is One, so all energy patterns are related to each other, a phenomenon we experience as causality. It is actually the divisions of the Formative

Mind which we experience as the divisions of the three Mind worlds. This is expressed at all levels of mind function: 'conscious', conceptual, subconscious, formative or causal. Or as the Eastern yogic aphorism puts it: '*Mind is the slayer of the Real.*'

In the physical, and indeed astral and causal worlds, the differing nature of these forms gives rise to souls encapsulated in different mind structures or patterns, which project outwardly as bodies. When we see only the bodies, we are confused. When we begin to understand the hidden laws of Mind function, then we begin to see things differently. When, at last, the soul stands naked in the spiritual regions, beyond the realm of the Universal Mind, then we know our real selves.

But this is beyond the reach of most forms of meditation. Only a perfect mystic can lift a soul to this height. The mystic experience of most folk in this physical universe, however illuminating, blissful and unifying they may be are usually only experiences within the subtle area in the 'sky' of our physical body, just below or occasionally penetrating into the subastral domains. This does not demean such experiences, for they are of tremendous importance to the individual, but it does tell us of the sheer immensity of the creation and of the great 'journey' the soul must travel to reach its original Home. Such experiences are only the beginning.

ENERGY AND THE FORMATIVE MIND

We often hear people speak of body, mind and soul and the words are used to convey many shades of meaning. But what are these 'entities'? And what is energy? What are substance and matter? Let us summarize it once again. Our soul is our consciousness, the innermost spark of life within us. It is a part of the Formless One, the One Ocean of Being. Consciousness is therefore One and *undivided.* No division means that it is form-*less.* That there is no difference within it, no pattern, no duality or polarity. All is One, an Ocean of Love and Life. This is the Source, the Uncreated God, Absolute Reality. It lies at the heart of what we are.

'Mind is the slayer of the Real.' Mind is a part of the process of creation. In the region of Universal Mind first arises duality or polarity – difference. From duality arises *division, form* or *pattern* and, simultaneously, as a part of the same thing, *space* and *time.* Space, time and pattern all imply division or difference, they are patterns of separation.

The Reality of pure consciousness is now overlain by Mind, though Mind has no independent existence, but arises due to the creative urge of Consciousness. This creative urge we call the Word, the Logos, the Creative Life Stream. It is also called the Sound Current or the Audible Life Stream, for this vibration of life can be *heard* within in mystic practice. It is actually the central pivot of the highest mystic teaching. The gospel of St John even starts out with this, 'In the beginning was the Word.' And we find the same central position given to this power of Life in the teachings of all mystics of the highest order, though there are also many imitations, too, which use the same language to describe lower regions of the grand hierarchy.

Through the causal, astral and physical worlds, the Formative Mind patterns become increasingly complex. All forms are thus of the Mind and we can look at our physical body as an image projected by consciousness though the screen of the Formative Mind. Consciousness shines through the Mind patterns and we perceive a physical body.

But actually, in the fuller sense, the patterns we call a physical body are actually a part of this ever-increasing complexity in the Mind's activity. *The physical body is thus an aspect of the Mind, itself.*

Similarly, the thoughts, emotions, intuitions and instincts which comprise what we normally think of as our human mind are only a small part of the totality of Mind. Mind is far more than our human experience of it. It consists of *everything* within our experience and much more besides.

To repeat the classification of terminology, given in Chapter 1, *Universal Mind* refers to the first beginnings of the Mind as it begins to weave its first patterns over the Creative Word. This is the area of the causal region. The *human mind*, spelt here with a small 'm', represents our individual physical mind, responsible for the formation of the patterns we call our destiny, as well as all the aspects of mind we know of in our life as humans – personality, thought, emotion, memory, intellect and much more. The *Formative Mind* represents the totality of all the aspects of the Mind – human and otherwise. It is the big egg of the Mind. The term '*Mind*' itself, with a capital 'M' represents the same. Our human mind, as well as the Universal Mind and indeed all its aspects – these are all formative, for this is its essential nature.

Now we can begin to see how everything fits together so appropriately, for everything is a part of the same Universal Formative Mind. Mind is more holistic, holographic, whole and integrated than one could ever imagine. And the underlying, cohering reality is of the One, of consciousness. Thus, all aspects of the Formative Mind patterns are related, because their substratum is One. This relationship among the Many, we call *causality*. It is the law of karma. It is the natural result of the creative presence of the One within the Many.

What, then, is *energy*? What we perceive as energy is nothing more than difference or pattern within the One. The creation is itself energy. Even above the level of the Universal Mind, there are spiritual regions created by the first, primal outpouring of the Creative Word. These are *created* regions. Only the undifferentiated One is beyond creation. So these primal, spiritual regions represent the first manifestation of creation, of difference within the One, that is – of energy patterns.

The simplicity of these primal energy currents or forms is multiplied as the Mind comes into action – or, more correctly – Mind is formed as the natural outcome of the multiplication of the primal energy currents, the primal creative urge, the primal form or difference.

It is because energy as we experience it is no more than these patterns of the Formative Mind spun across the inward power of the One, that so many people feel instinctively that the *energy paradigm* is so fundamental and all-embracing. 'Everything is energy' is a saying one hears so very frequently. We perceive energy as the One in action, the One become the Many, divided by Mind.

In this physical universe and from a pragmatic point of view, considering our complete involvement with Mind, this appears to be true, though, as we have seen, there are even primal energy patterns, beyond the Formative Mind, in the purely spiritual realms of creation. But the true spark of consciousness, of life, is the primal undifferentiated energy or Life Force itself.

Within the Mind worlds, we perceive energy as the patterns of creation, of differentiation. So fine and tenuous a net is woven that the Mind seems to be the creator and the prime cause. It is because of this, that some mystics have mistaken the Universal Mind for the Supreme Being Himself. Then they talk of the Divine Mind, describing the primal 'divine' law as that of justice, karma or causality. This is what tells us just which region and which 'Lord' it is that is being described. For the law of the Supreme is of Love, merging and Oneness. The god of justice is the Universal Mind.

Incidentally, this is the origin of the concept of Satan, derived from the Arabic word Shaitan, or the Hebrew, Satan. Satan is the Universal Mind, the great power that administers the three regions of the Mind.

So the Mind itself is the 'devil', the slayer of the real, though created by the Supreme and acting within His Will. Indeed, in these higher worlds of the Mind, the 'devil' is fascinatingly and blissfully beguiling. 'He' is not all fire and brimstone! Mostly the reverse, in fact. And 'he' never acts outside the jurisdiction of the One.

Consciousness, when seen in its purified essence, is One, undivided. It is primal, undifferentiated power or energy. It is also the sustainer and creator of all energy patterns, like the ocean supports the waves, yet the waves are never separate from the ocean. But fundamentally and in essence, consciousness is formless and uncreated, a drop of the Supreme Being.

Energy is thus both the Creator and the Creation. In the Mind worlds, energy is the patterning of that primal power. All patterns are thus of the Mind. But the power within it, which gives it existence, is the One Ocean of Consciousness, the Creative Life Stream. Or in simple terms – there is just one Big Thing going on!

THE WHOLENESS OF THE ENERGY DANCE

This process of forming increasingly complex patterns from within-out gives us a most amazingly breathtaking and beguiling spectacle of integration and wholeness, for the One is always present within everything. The more inward patterns, the more inward energies, are formed into the more outward and more diverse patterns, but the laws governing this multi-reflective, multi-faceted, multi-layered, multi-dimensional process are way beyond the wit of our human intellect to grasp. Our human mind and intellect are themselves only a part of the total pattern of the Formative Mind.

With this vision or understanding of the symphonic, multilevel wholeness of the Mind, we can begin to glimpse how our physical universe and physical body are constructed. But still, we are only looking at the outer layers of this incredible cosmic onion. So we see only effects.

We can intuitively perceive the relevance in coming to grips with the fact that all sensory experience is subjective, incommunicable and in our own minds – that the apparently outward world is thus within our own mind structure. With the body as only an outer shell of the Formative Mind's patterning process, we can understand the integral blueprinting mechanisms by which it is formed or patterned and how our whole existence is really psychosomatic – mind and body are part and parcel of the same process of Mind.

We can see how the body-emotion-mind structure can be seen as patterns and relationships of energy, for pattern and relationship are all that the creation is – this is the nature of form.

We can see how spiritual evolution entails the shifting of one's point of perception ever higher or more inward within the integrated patternings of the Mind. And it becomes clear why spiritual progress requires meditation, control of the mind. And why the highest form of meditation is to ride on the backbone, the spine, the substratum, the Creative Essence of all the Mind's diversity, that is, to contact the underlying Creative and Audible Life Stream itself.

With the physical universe being described as no larger than a hair in an ocean when seen within the context of the more inward realms of the Mind, one can see too why there are so many schools of meditation. For meditation has come to mean developing contact with any part of our existence other than the gross physical, shutting one's physical eyes and doing almost anything within oneself. This includes contacting the subtle patterns or energies of our physical covering – the pranas, chakras and physical tattwas – or enlarging consciousness of our human mind structure. Even just sitting and cogitating has come to be called meditation. There is no lack of inward things to explore.

Automatically, too, as their own mind becomes more subtle, practitioners of meditation become more aware of the subtle patterns of existence. Clairvoyance, psychism, telepathy, even the ability to heal or perform miracles automatically comes to such practitioners as they begin to perceive, intuitively and almost unconsciously at first, how the patterns are put together, related and integrated.

But then egotism can come into play and the soul can remain entranced with such phenomena for a long time, while further progress ceases. The attention plays on the higher floors of this multilevel department store of the Mind, but in so doing, forgets the central lift shaft of the Creative Word and the One who dwells in the Penthouse Suite!

So it is all one big dance of creation, of energy, of pattern, spun out by the Formative Mind. In this, we have danced for innumerable lives – sometimes as an ant or a plant, sometimes as a bird or beast, sometimes as a human, sometimes as a temporary denizen of the higher realms of the astral, perhaps even the causal, realms. But always we have danced to the tune of the Mind, and will continue to do so, evolving and devolving on the inward ladder of consciousness, until we are released from this great jailhouse of creation. And this can only happen by the Will of the Creator Himself.

EPILOGUE

A PLACE FOR EVERYTHING

The beauty of the proposal I am putting forward is that it does not prove any other theory to be completely incorrect. It just adds in a new dimension, a continuously creative energy projection dimension of Mind and consciousness, which provides understanding of so many factors in science, in human psychology and in everyday experience, which puzzle us all.

It is as if we had lived in a two-dimensional world all our life and never noticed the existence of a real, third spatial dimension. When this third dimension is pointed out to us it explains so many phenomena in such a simple fashion that our first instincts are to reject it, to return to our more familiar restrictive world of anomaly and paradox.

But this description of the Formative Mind and Consciousness and its periodicity in subtlety and physical expression, demands to be taken seriously. For no other description of the universe covers so much ground. There are many, many details to be filled in, but in one all-encompassing swoop it tells us:

1. The origin of life and consciousness.
2. The origin of all forms and patterns.
3. The origin and nature of causality and energy.
4. The nature and inner structure of man.
5. The nature and inner structure of all other creatures.
6. The nature and status of all man's science – the reasons why we perceive repeating patterns in nature and why we find geometric, mathematical and statistical 'laws of nature'.
7. Why 'conventional' modern physics seeks for a unity it has not yet found.
8. The nature of mind, matter, energy, the 'laws of nature' and consciousness.

9. The links between mind, emotion, body, psychotherapy and brain physiology/anatomy.

10. An understanding of neurosis, psychosis, mental disturbance and various other mental states and conditions of consciousness.

11. An understanding of the dynamic energy structuring of the human mind and aspects of the greater Mind.

12. The way in which man differs from other species and how they differ from each other.

13. The manner by which the 'evolutionary' process (actually, a process of change, rather than evolution) has come about.

14. How many of the 'anomalies' of the fossil record have come into being, e.g. the absence of 'missing links' between species and the origin of *punctuated equilibria.*

15. The way in which outward events relate to the content of our minds.

16. The manner by which coincidences and synchronicity come about.

17. The way in which all psychic and paranormal phenomena come about.

18. The nature of mystic and religious experience.

19. The reason why religions hold such a power over human minds; why they all have the same essential spiritual and moral basis; how religions arise.

20. Why man even rejects the very idea of a hidden unity and deeper processes of the mind.

21. Why holistic principles are so appealing, and the deeper, inward origin of such principles.

22. Why the universe is put together in such a way that it has been called *The Symbiotic Universe,* and how the apparently amazing universal coincidences leading to the formulation of this *Anthropic Principle* have actually come into existence.

23. How the ordered maintenance of the planetary biosphere by the living creatures themselves, as described in the *Gaia Hypothesis* of James Lovelock is both real and quite understandable, even predictable.

24. How Lyall Watson's observations of a *Supernature;* how the nested and formative patterns perceived and described in such excellent detail by Rupert Sheldrake; how the *Collective Unconscious* and the unconscious archetypes of Carl Jung; how the theories of a self-organizing universe — how all these are only aspects or outward perceptions and observations of the

inwardly dimensioned and wholly integrated natural creation spun out by the Formative Mind.

These, and all other aspects of mind and physical experience can *all* be understood within this framework, for it tells us of how the creation is put together, of the relative position of all things.

And it is not a dogma, nor even a philosophy in the normal sense of the term, for ultimately it points to an empirical experience. It acknowledges that nothing exists outside our subjective experience, not even our sensory experience; and that ultimately the entire structure of creation, and the Creator Himself, can be directly perceived by a process of experiential knowing, an expansion of consciousness.

Just as we *know* we exist without needing to prove it to anyone, just so can we know the higher Truth of existence in inward mystical experience, automatically recognizing the experience as real, without ever needing to justify the validity of the experience. Then we know; we do not surmise.

Moreover, this description of the nature of life is nothing new. The essence is the same mystical philosophy which is found in almost all cultures, often obscured by our human confusions, but present nonetheless. All the above topics are considered in my series of six books concerning mysticism, science and the natural world. In them I have used modern language and modern patterns of thought to point out the position of modern science in relation to this ancient philosophy. My writing may therefore have a ring of newness. But the essence is as old as the creation. This is the natural philosophy of the creation, not something spun out of the confused human mind.

In many respects it can be honestly said that I have only put some modern flesh and detail onto a feeling common to many people, that the universe is the 'body' or projection of God, manifesting His intricate design and purpose. Again, I am fond of pointing out that the universe is both a continuous show as well as a multilevel affair. But this is also a perception common to almost all esoteric and many religious schools of thought.

I have also written many times that the real answers only come from inner mystic experience. Yet over two thousand years ago, Greek mystics pointed out that one could know everything by looking within oneself – though special techniques are required. Later Greek sophists and more modern intellectual philosophers have misunderstood this to mean that everything could be understood simply by *thinking* about it. But this is a far cry from the real meaning of those ancient mystics. What they meant was that everything is within us and can be known and understood from within in sublime mystic experience. Thinking and true mystic experience are quite different ways of knowing about things.

WHERE NEXT?

Natural Creation and the Formative Mind is the first in this series of three books: *Natural Creation . . . Or Natural Selection?* and *Natural Creation: The Mystic Harmony* are the last two.

In *Natural Creation and the Formative Mind* I have taken the perennial, mystic philosophy and applied its perceptions to the world of physical creatures. This identifies very clearly how it is that we can feel a kinship with all life forms, for the deepest spark of life within us all is the same. We are all 'hiddenly linked'. Similarly, although our minds and bodies display certain differences, yet the essential creative process, the principle of Formative Mind function, is the same throughout. Therefore we are aware of other creatures, perceive them with our senses, and can relate to them, though we may not fully understand them. For we are all members of a greater family with no really alien forms among us.

In *Natural Creation . . . Or Natural Selection?* these insights are applied to the greater cycles through which our planet passes, going back hundreds of millions of years. And we find that these too, are patterns of the Formative Mind. So we look at the fossil record, the remains of old bodies from which the mind and soul have long since fled. Yet those souls, taking birth continuously, under the influence of their minds, have remained. We find them as the creatures of our present times. We are those souls. And we also look at the ancient geological formations, the inert and rocky remains of the physical arena in which those ancient minds, bodies and life forms had their day.

Most importantly, we discuss in some considerable detail the implications and the evidence for the greatest cycle of all which lies within Mind and Consciousness itself, the cycle of the Life Force.

These deeply relevant perceptions of the inner and hidden dimension to life are also applied to Darwin's Victorian theory of evolution, as it is presently interpreted by material science. And we find that a most remarkable and fascinating picture emerges. One that both completely and rationally explains the nature and content of the fossil record and yet places it within the context of a process of continuous manifestation and an understanding of Mind and Consciousness as the prime movers in creation.

Finally, *Natural Creation: The Mystic Harmony* describes the intricate ordering of the physical universe, which modern science attributes to a mixture of self-organization and coincidence. It shows how this order comes into being as projected patterns of the Formative Mind. It gives many examples of the mystic experience, as well as the incidence of the same old mystic philosophy among the lesser known cultures of the world. And it ends with a description of the grand inner hierarchy of creation as given by all true mystics.

BIBLIOGRAPHY

Books quoted or from which material has been drawn are all listed below. A few others of related interest are also included.

Animals and Other Species

Backster, Cleve, *Evidence of Primary Perception in Plant Life*; International Journal of Parapsychology 10, 1968.

Bonner, W. Nigel, *Whales*; Blandford Press, 1980.

Boone, J. Allen, *Kinship with all Life*; Harper and Row, 1954.

Callahan, Philip, *Tuning in to Nature*; Devin-Adair, 1975.

Corbett, Jim, *Man-Eaters of Kumaon*; Penguin, 1944.

Downer, John, *Supersense*; BBC Books, 1988.

Fenton, M. Brock, *Communication in the Chiroptera*; Indiana University Press, 1985.

Hansell, Michael, *Animal Architecture and Building Behaviour*; Longman, 1984.

Lawick-Goodhall, Jan van, *In the Shadow of Man*; Collins, 1971.

Lorenz, Konrad, *King Solomon's Ring*; Methuen, 1952.

Norris, K.S., *The Echolocation of Marine Mammals* (1969); in *The Biology of Marine Mammals*, edited H.T. Anderson; Academic Press.

Norris, K.S., *Some Problems of Echolocation in Cetaceans* (1964); in *Marine Bio-Acoustics*, edited by W.N. Tavolga, Pergamon Press.

Savage-Rumbaugh, E. Sue, *Ape Language*; Columbia University Press, 1986.

Tompkins, Peter and Christopher Bird, *The Secret Life of Plants*; Harper and Row, 1973.

Watson, Lyall, *Supernature*; Hodder and Stoughton, 1973.

Miscellaneous

Beston, Henry, *The Outermost House*; Selwyn and Blount, 1928.

Murphy, Michael and Rhea White, *The Psychic Side of Sport*; Addison-Wesley, 1978.
van der Post, Laurens, with Jean-Marc Pottiez, *Walk with a White Bushman*; Chatto and Windus, 1986.

Mysticism and Mystical Experience

Gollancz, Victor, *A New Year of Grace*; Victor Gollancz, 1961.
James, Joseph, *The Way of Mysticism*; Jonathan Cape, 1950.
Jeffries, Richard, *The Story of My Heart*; Longman, Green & Co., 1883.
Johnson, Raynor C., *The Watcher on the Hills*; Hodder and Stoughton, 1959.
Johnson, Raynor C., *The Imprisoned Splendour*; Hodder and Stoughton, 1953.
Maharaj Charan Singh Ji, *The Master Answers*; Radha Soami Satsang Beas (India), 1966.
Maharaj Sawan Singh Ji, *Philosophy of the Masters*; Radha Soami Satsang Beas (India), 1965.
Maharaj Sawan Singh Ji, *Spiritual Gems*; Radha Soami Satsang Beas (India), 1965.
Mingana, A, *Early Christian Mystics, Christian Documents in Syriac, Arabic and Garshuni*, edited and translated; W. Heffer and Sons, Cambridge, 1934.
Yogananda, Paramhansa, *Autobiography of a Yogi*; Hutchinson, 1950.

Organization and Order in the Universe

Barrow, John D. and Frank J. Tipler, *The Anthropic Cosmological Principle*; Oxford University Press, 1979.
Greenstein, George, *The Symbiotic Universe*; William Morrow and Company, 1988.
Jantsch, Erich, *The Self-Organizing Universe*; Pergamon Press, 1980.
Lawlor, Robert, *Sacred Geometry*; Thames and Hudson, 1982.
Lovelock, James, *The Ages of Gaia*; Oxford University Press, 1988.
Lovelock, James, *Gaia*; Oxford University Press, 1979.
Prigogine, Ilya and Isabelle Stengers, *Order Out of Chaos*; William Heinemann, 1984.
Sheldrake, Rupert, *A New Science of Life, The Hypothesis of Formative Causation*; Blond and Briggs, 1981.
Sheldrake, Rupert, *The Presence of the Past*; Collins, 1988.

GLOSSARY

Anda   Literally, 'egg'. The astral realm lying between the causal region (above and Pinda, the physical universe, (below). The heavens spoken of by most religions lie within Anda.

astral region   *See* Anda

Aum   The Creative Word of the Universal Mind region. Alternative spelling: Om. *See also* Life Force

Brahm   'Lord' or 'ruler' of the region of Universal Mind.

Brahmanda   Literally, 'Egg of Brahm'. Refers either to the causal region of Universal Mind, or to the regions of Anda (the astral realms) and Brahmanda combined.

causal region   *See* Brahmanda

cerebral cortex   The outer layers of the brain, largely responsible for sensory and motor perception and co-ordination.

cetaceans   An order of aquatic mammals, possessing no hind limbs and a blow-hole for breathing. For example, whales, porpoises, dolphins etc.

chakras   Centres of pranic energy and organization within the physical body. There are five such centres, one related to each of the five tattwas, with a sixth 'controlling' centre – the eye centre. Some western schools talk of a seventh, 'crown' chakra, but this is really a confusion, for it relates to the Sahasra, which is not a part of Pinda, the physical universe, at all. At death, the pranas are withdrawn and the chakras are dissolved. Sahasra, however, is not affected by bodily death.

chaurasi   The wheel of 8,400,000 species or bodily forms in which a soul may take birth whilst in the physical world.

Creative Word   *See* Life Force

Daswan Dwar   Literally, 'tenth door'. The first region of pure spirit, lying immediately above the Universal Mind. It is here that the soul first knows itself as pure soul. This high mystic experience is the true self-realization. Daswan Dwar is so called because the region of the Universal Mind is said to have nine 'passages', 'currents' or 'doorways' opening downwards towards the lower creation, with only one leading upwards – the tenth door.

DNA   Deoxy-ribose-nucleic acid. The main constituent of the chromosomes of all living organisms (except some viruses), playing an important role in protein synthesis and in the transmission of hereditary characteristics.

duality   The physical universe is said to be the world of duality, of the pairs of opposites: hot and cold, truth and untruth, positive and negative, activity and inertia – and so on. Such division arises in seed or essence form in the Universal Mind and finds its most crystallized expression or manifestation, here on the physical plane.

Egg of Brahm   *See* Brahmanda

endocrine glands   Glands which produce hormones. *See also:* hormones

eye centre   The sixth chakra of Pinda, lying between and behind the two physical eyes, though it has no specific anatomical location. By concentration at this point, with the suitable guidance and inner help of a guru, a disciple is able to withdraw all attention and consciousness from the physical body and go into the inner realms of creation in mystic transport. He thus passes through the process of death whilst still living in the physical body. Sometimes also called the third eye.

Feng Shui   Literally, 'wind and water'. The Chinese art of arranging the environment, both indoors and outside, in order to best adjust its subtle energy flow (Ch'i), the idea being to create a healthy, prosperous and harmonious atmosphere.

gunas   The three primary modes or attributes to be found in all the Mind worlds. First arising in the region of the Universal Mind and running throughout all creation below, they are (for example): creation, preservation, destruction; future, present, past; birth, life, death; outgoing balance, indrawing; activity, harmony, inertia; positive, zero, negative – and so on.

Guru   Literally, 'one who gives light'. A spiritual teacher; a mystic teacher; one who teaches the practical techniques of spiritual and mystic progress; a Master. A perfect Guru (Satguru) or Perfect Master is one who comes from the Supreme Ocean of Being with the intention of releasing certain souls from the wheel of birth and death. He is one with the Creative Word – he is the 'Word made flesh.' In fact, the real Master is not the body, but the Word. There is always at least one Perfect Master alive at any time. *See also:* karma

hormones   Biochemical substances produced by particular organs (eg. endocrine glands) or tissues within living organisms which have specific effects on tissues at a different location within that body.

indriyas   The sense and motor organs, there being five of each. To each sensory and motor function there are two aspects: the physical organ and its mental counterpart. In this book, as in some other yogic texts, the indryas are taken to refer to the mental counterpart only.

Kal   The Negative Power; Satan; the lord or ruler of the Universal Mind; Brahm.

karma   The law of the great Mind by which souls, under the influence of their individual minds are brought back into physical birth after physical death. It is an automatic and natural law of cause and effect which exacts strict justice rather than displaying mercy and forgiveness. The record of karma is held within the Mind as the impressions in seed form of all thoughts, actions and desires ever entertained or performed by a soul. Karma is of three kinds:

1. **Destiny karma** – the events of life which are fixed at the time of birth and which have to be undergone. They are the effects, good and bad, of previous actions, thoughts and desires from previous lives.

Destiny is etched into the complex, energetic fabric of our human mind.

2. New karma – new actions and desires, performed or entertained in the present life, which become seeds or mental impressions for the destiny of future lives.

3. Stored karma – in one lifetime, we may gather more new karma than can be paid off in just one future life. Any balance of this 'unused' karma goes into 'storage'. Over the span of aeons, this store of karma becomes a great weight upon the soul, keeping it bound to the wheel of birth and death. This continues indefinitely, until the soul has the good fortune to meet a Master who has come from beyond the realm of karma and has the power to release souls from the Mind, by taking on responsibility for the payment of this vast debt.

Karma also means 'sins' and, from a mystic point of view, the 'forgiveness of sins' actually means help and guidance in the clearing of this great mountain of stored karma by a contemporary, living Master who is divinely qualified and appointed to do so. Such a Master inwardly connects or re-tunes the soul to the Creative Word or Logos – which is the real Master. It is the absorption of the mind and soul in the sound and light of the Creative Word which cleanses it of all past karmas and ultimately draws it back to God, beyond the realm of Mind and Maya. *See also* Guru.

Life Force   This term is used in many ways by different writers. In this book, it refers to the creative power of God in the creation; His primal emanation, from which all other emanations are derived. It is the primal vibration, underlying all other vibration and movement in the creation, both within and without. Nothing exists, at any level in creation, without this incessant movement. And just as our senses can perceive certain physical vibrations, so too do our inner mind and soul possess the capacity to *hear* this Primal Vibration inside, in mystic experience. The inner light also comes from this Primal Sound. Also called: Logos, Word, Creative Word, Sound Current, Audible Life Stream, Life Stream. This life-giving current, as heard in the region of the Universal Mind, is called Aum or Om, in the Vedas.

Logos   *See* Life Force

Maya   Literally, 'illusion'. First arises in the region of the Universal Mind as a web of patterns and rhythms over the face of pure spirit. It is illusory because the mind takes these patterns and rhythms as the reality, losing sight of the One Creator within all things. The sensory world of the physical universe as well as the higher regions of the Mind are thus all Maya – illusion.

mid-brain   The central portions of the brain.

pheromones   Biochemical substances secreted by many creatures as a means of communication, mostly between members of the same species.

Pinda   The physical universe, including both its gross and subtle aspects; also the physical body, the microcosm through which the physical universe is both interacted with, and manifested.

prana   The Life Force, stepped down – so to speak – in order to energize the physical body, thereby giving it life. The subtle vibration or energy of life which orders and integrates the amazing mozaic of biochemical and bodily processes.

Puranas   Literally, the 'old ones'. Any of a class of ancient Sanskrit writings not included in the Vedas, said to have been compiled by the rishi (yogi or sage) Vyas.

radar   A means of detecting the position and speed of objects. A narrow beam of electromagnetic pulses is transmitted and reflected back from surrounding objects, giving information on those objects. Human instrumentation uses electromagnetic pulses, but many creatures use sound pulses. Hence the term 'acoustic radar' – also known as sonar or echolocation.

Sach Khand   Literally, 'true home'. The eternal region of the soul, not subject to dissolution.

Sahans dal Kanwall   *See* Sahasra

Sahasra   The central powerhouse of Anda, the astral region, supporting all creation below. Also called: Thousand-Petalled Lotus and Sahans dal Kanwal.

Sanskrit   An ancient language of India; the language of the Vedas, of Hinduism, and of an extensive scientific and philosophical literature dating from the first millennium BC. The oldest recorded language of the Indic branch of the Indo-European group of languages, recognized as such in the eighteenth century from a comparison of Greek and Latin with Sanskrit.

sat Literally, 'truth', specifically Divine truth; that which endures, rather than that which changes and is subject to dissolution.

steroids A large group of fat soluble biochemical substances characterized by a chemical ring structure. The majority, such as D vitamins and many hormones have important physiological effects.

Tattwas Energy essences, primary elements. First manifesting as ultra-fine mental essences, blueprints or energies in the region of the Universal Mind, the tattwas appear as reflections at the physical level and are involved in all aspects of physical living, gross and subtle.

Thousand-Petalled Lotus So called because it consists of 1000 energy currents, derived from a multiplexing of the energy currents which feed it from above. *See* Sahasra.

Vedas Literally, 'divine knowledge'. The four principal holy books of the Hindus, written by their great sages and mystics in ancient times.

Upanishads Literally, 'to sit near or close to' or 'setting at rest ignorance by revealing the knowledge of the Supreme Spirit'. Thus it also means, esoteric doctrine or mystical teachings. Specifically, the Upanishads are a class of Hindu writings, there being 108 in number, whose aim is to reveal the secret or mystical meaning of the Vedas.

yin and **yang** The duality of the physical universe; the pairs of opposites, according to Chinese philosophy. Yin refers to the receptive, negative, feminine polarity; yang to the expressive, positive, male polarity. *See also*: duality.

yugas Literally, 'ages'. According to the Hindu sages, the physical universe is said to cycle through ages of great spirituality (Sat yugas) down to ages of great spiritual poverty (Kal yugas). The span of one cycle lasts 4.32 million years. Just as spring comes after winter, a Kal yuya is followed by a Sat yuga. We are presently passing through Kal yuga.

FURTHER INFORMATION

Details of further books by the same author, as well as books upon allied topics, are available from:

> WHOLISTIC RESEARCH COMPANY
> Bright Haven,
> Robin's Lane,
> Lolworth,
> Cambridge CB3 8HH,
> England.

Please send four second class postage stamps in the U.K. (£2.50 or $5.00 for overseas) for a booklist.

Visits by appointment only, please

INDEX